Numbers, sets and axioms
THE APPARATUS OF MATHEMATICS

A.G.HAMILTON
Lecturer in mathematics at the University of Stirling

CAMBRIDGE UNIVERSITY PRESS

Cambridge
London New York New Rochelle
Melbourne Sydney

Published by the Press Syndicate of the University of Cambridge
The Pitt Building, Trumpington Street, Cambridge CB2 1RP
32 East 57th Street, New York, NY 10022, USA
296 Beaconsfield Parade, Middle Park, Melbourne 3206, Australia

First published 1982

Printed in Great Britain
at the University Press, Cambridge

Library of Congress catalogue card number: 82-4206

British Library cataloguing in publication data

Hamilton, A. G.
Numbers, sets and axioms: the apparatus of mathematics.

I. Set theory
I. Title
511.33   QA248

ISBN 0 521 24509 5 hard covers
ISBN 0 521 28761 8 paperback

CONTENTS

PREFACE

The mathematician's work proceeds in two directions: outwards and inwards. Mathematical research is constantly seeking to pursue consequences of earlier work and to postulate new sorts of entities, seeking to demonstrate that they have consistent and useful properties. And on the other hand, part of mathematics consists of introspection, of a process backwards in the logical sequence, of the study of the nature and the basis of the subject itself. This inward-looking part is what is normally called the foundations of mathematics, and it includes study of set theory, the number systems, (at least some) mathematical logic, and the history and philosophy of mathematics. This book is not intended to cover all of these areas comprehensively. It is intended to convey an impression of what the foundations of mathematics are, and to contain accessible information about the fundamental conceptual and formal apparatus that the present-day working mathematician relies upon. (This apt description is due to G. T. Kneebone.) The concepts directly involved are: numbers (the various number systems), sets, orderings of sets, abstract mathematical structures, axiomatic systems, and cardinal and ordinal numbers.

The book presupposes no knowledge of mathematical logic. It does presuppose a certain amount of experience with mathematical ideas, in particular the algebra of sets and the beginnings of mathematical analysis and abstract algebra. Its structure is in many ways a compromise between what is appropriate for a course text and what is appropriate for a more readable background text. Different teachers would place different emphases on the topics here dealt with, and there is no clear logical sequence amongst the topics. It is intended, therefore, that the chapters (at least the earlier ones) should have few interdependences. There are forward and backward references where appropriate, but these are generally illustrative rather than essential. Each chapter begins with a brief summary which includes an indication of its relationships with the other chapters. The level of mathematical sophistication does increase as the book proceeds, but throughout the book there are exercises of a routine nature as well as more taxing ones. These exercises are an integral part of the presentation, and they are designed to help the reader to consolidate the material.

The content is definitely mathematics, not logic or philosophy. Nevertheless, no book on this sort of subject matter can avoid involvement with philosophical issues. As readers will discover, underlying the whole of the presentation is a definite philosophical viewpoint, which may be summarised in the following remarks. Mathematics is based on human perceptions, and axiom systems for number theory or set theory represent no more than a measure of the agreement that exists (at any given time) amongst mathematicians about these perceptions. Too much prominence has been given to formal set theory as a foundation for mathematics and to sets as 'the' fundamental mathematical objects. Mathematics as a whole is *not* a formal axiomatic system. Mathematics as practised is not merely the deduction of theorems from axioms. Basing all of mathematics on set theory begs too many questions about the nature and properties of sets. Thus, while axiomatic set theory is treated in some detail in Chapter 4, it is treated as a *part* of mathematics, and not as a basis for it.

Following each chapter there is a list of references for further reading with some comments on each one. The details of these titles will be found in the full bibliography at the end of the book. Also provided at the end of the book are a glossary of symbols and hints and answers to selected exercises. In the text the symbol ► is used to denote the resumption of the main exposition after it has been broken by a theorem, example, remark, corollary or definition.

During the development of a text and its presentation it is very useful for an author to have the work read by another person. A more detached reader can spot deficiencies which, whether through over-familiarity or carelessness, are not apparent to the writer. I have been fortunate, in this respect, to have been kindly and conscientiously assisted by Dr G. T. Kneebone and Dr John Mayberry. Both have read several versions of the various chapters and have made numerous suggestions for improvement, most of which have been incorporated in the book. I am most grateful to them for their effort, and I can only hope that they will derive some satisfaction from the final version which is my response to their comments.

May Abrahamson made a marvellous job of typing (and patiently re-typing). Cambridge University Press has again been most efficient and helpful. The University of Stirling has assisted by allowing me leave during which the bulk of the writing was done. My sincere thanks go to all of these.

A.G.H.   January 1982

1

NUMBERS

Summary
First we consider what are the basic notions of mathematics, and emphasise the need for mathematicians to agree on a common starting point for their deductions. Peano's axioms for the natural numbers are listed. Starting with a system of numbers satisfying Peano's axioms, we construct by algebraic methods the systems of integers, rational numbers, real numbers and complex numbers. At each stage it is made clear what properties the system constructed has and how each number system is contained in the next one. In the last section there is a discussion of decimal representation of rational numbers and real numbers.

The reader is presumed to have some experience of working with sets and functions, and to be familiar with the ideas of bijection, equivalence relation and equivalence class.

1.1 **Natural numbers and integers**

It is fashionable nowadays at all levels of study from elementary school to university research, to regard the notion of set as the basic notion which underlies all of mathematics. The standpoint of this book is that the idea of set is something that no modern mathematician can be without, but that it is first and foremost a *tool* for the mathematician, a helpful way of dealing with mathematical entities and deductions. As such, of course, it becomes also an object of study by mathematics. It is inherent in the nature of mathematics that it includes the study of the methods used in the subject; this is the cause of much difficulty and misunderstanding, since it apparently involves a vicious circle. The trouble is that most people (mathematicians included) try to regard mathematics as a *whole* – a logical system for proving true theorems based on indubitable principles. The present author believes that this is a misleading picture. Mathematics is rather a mixture of intuition, analogy and logic – a body of accepted knowledge based on perceived reality, together with tools and techniques for drawing analogies, making conjectures and providing logical justification for conclusions drawn.

The fundamental notions of mathematics now are the same as they were a hundred years ago, namely, numbers or, to be more specific, the number systems. Modern abstract mathematics (with the exception, perhaps, of geometry and topology) is based almost entirely on analogies drawn with properties of numbers. Here are some simple examples. The algebraic theory of fields arises from a generalisation of the properties of addition and multiplication of numbers. Real analysis is just the study of functions from real numbers to real numbers. Functional analysis applies the methods of algebra (themselves derived from methods used in concrete numerical situations) to mathematical systems which are generalisations of three-dimensional physical space (which can be represented, of course, via coordinate geometry, by means of ordered triples of real numbers).

Our knowledge of the number systems derives from our perception of the physical world. We count and we measure, and the origins of mathematics lie in these activities. Modern methods can help in writing down and working out properties of numbers and in clarifying relationships between these properties. Indeed, this process has reached an advanced stage. Most mathematicians now agree on what are the principles which it is proper to use in order to characterise the number systems. This is very significant, for it provides a common starting point for logical deductions. If all mathematicians based their deductions on their own personal intuitions then communication would be very difficult and the subject would not be very coherent. One purpose of this book is to expound and explain the common starting point. In the first chapter we deal with the number systems out of which mathematics develops, and in subsequent chapters we shall investigate some of the tools (notably set theory) and try to explain what modern 'foundations of mathematics' is all about.

Counting is the first mathematical activity that we learn. We learn to associate the objects in a collection with words (numbers) which mark them off in a sequence and finally indicate 'how many' there are in the collection. This experience gives us an intuition about an unending sequence of numbers which can be used to count in this way any finite collection of objects. It is assumed that the readers of this book will have a well-developed intuition about natural numbers, so we shall not go into the psychology behind it (this is not to imply that the psychology of mathematical intuition is not worthy of study – just that it is outside the scope of this book).

Notation The set of natural numbers will be denoted by $\mathbb{N}$. $\mathbb{N}$ is the collection $\{0, 1, 2, \ldots\}$. Notice that we include 0 in $\mathbb{N}$. This is merely a convention. It is common but not universal.

Let us list some properties of these numbers which accord with intuition.

Examples 1.1
(a) There is an addition operation (a two-place function on $\mathbb{N}$) which is commutative and associative.
(b) $0 + n = n$, for every $n \in \mathbb{N}$.
(c) In the list $\{0, 1, 2, \ldots\}$, the number following n is $n + 1$, for each n.
(d) $n + 1 \neq n$, for every $n \in \mathbb{N}$.
(e) $m + 1 = n + 1$ implies $m = n$, for every $m, n \in \mathbb{N}$.
(f) There is a multiplication operation (also a two-place function on $\mathbb{N}$) which is commutative and associative, and which distributes over addition.
(g) $0 \times n = 0$, and $1 \times n = n$, for every $n \in \mathbb{N}$.
(h) $m \times n = p \times n$ implies $m = p$, for every $m, n, p \in \mathbb{N}$ $(n \neq 0)$.

▶ Clearly, we can continue writing down such properties indefinitely. These are the kind of things we learn in elementary school. We learn them, discover that they work, and come to believe them as truths which do not require justification. However, the mathematician who is working in the theory of numbers needs a starting point in common with other mathematicians. *Peano's axioms* (listed first in 1888 by Dedekind, and not originating with Peano) are such a common starting point. They are five basic properties, all of them intuitively true, which serve as a basis for logical deduction of true theorems about numbers. They are as follows.

(P1) There is a number 0.
(P2) For each number n, there is another number n' (the *successor* of n).
(P3) For no number n is n' equal to 0.

(P4) If m and n are numbers and $m' = n'$, then $m = n$.

(P5) If A is a set of numbers which contains 0 and contains n' for every $n \in A$, then A contains all numbers.

(Note that we have used the word 'number' here as an abbreviation for 'natural number'.)

Remarks 1.2

(a) (P1) and (P2) provide the process for generating the sequence of natural numbers corresponding to the intuitive counting procedure. (P3) reflects the fact that the sequence has a beginning.

(b) (P4) is a more complicated property of the sequence of numbers: different numbers have different successors.

(c) (P5) is the principle of mathematical induction. This is the most substantial of the five, and is the basis of most proofs in elementary number theory. It may be more familiar as a method of proof rather than an axiom, and in a slightly different form: if $P(n)$ is a statement about a natural number n such that $P(0)$ holds, and $P(k+1)$ holds whenever $P(k)$ holds, then $P(n)$ holds for every natural number n. This can be seen to be equivalent to (P5) if we think of the set A and the statement $P(n)$ related by:

$n \in A$ if and only if $P(n)$ holds.

Thus, given a set A, a statement $P(n)$ is determined and vice versa.

▶ There is no mention in Peano's axioms of the operations of addition and multiplication. This is because these can be defined in terms of the other notions present.

The common starting point, therefore, *need not* mention these operations. However, it is quite difficult to carry out the procedure of defining them and justifying their existence (see Section 4.3) and, for our present purposes, it is certainly unnecessary. For our purposes we can broaden the common starting point, that is to say, we can include amongst our basic intuitive properties the following.

(A)   There is a two-place function (denoted by +) with the properties:

$m + 0 = m$,   for every number m.

$m + n' = (m + n)'$,   for all numbers m, n.

(M)   There is a two-place function (denoted by ×) with the properties:

$m \times 0 = 0$,   for every number m.

$m \times n' = (m \times n) + m$,   for all numbers m, n.

(Following the usual mathematical practice we shall usually omit multiplication signs, and write mn rather than $m \times n$. The exceptions will be when special emphasis is being placed on the operation of multiplication.)

We could also include amongst our basic intuitive properties the assertions that these operations satisfy the commutative, associative and distributive laws, but it is not difficult to prove, from the properties given above, that these hold. Let us carry out one such proof, as an example.

Theorem 1.3
Addition on $\mathbb{N}$ is commutative.

Proof
This is an exercise in proof by induction. We require two preliminary results:

(i) $0 + m = m$,   for all $m \in \mathbb{N}$.

(ii) $m' + n = (m + n)'$,   for all m, $n \in \mathbb{N}$.

For (i), we use induction on m. By property (A) we have $0 + 0 = 0$. Suppose that $0 + k = k$. Then $0 + k' = (0 + k)' = k'$ (using property (A) again).

Hence, by the induction principle, $0 + m = m$ holds for all $m \in \mathbb{N}$.

For (ii), we apply (P5) to the set

$$A = \{n \in \mathbb{N}: m' + n = (m + n)', \text{for every } m \in \mathbb{N}\}.$$

First, $0 \in A$, since $m' + 0 = m'$ (by property (A)) and $(m + 0)' = (m)'$ (again by property (A)), and so $m' + 0 = (m + 0)'$, for any $m \in \mathbb{N}$. Second, suppose that $k \in A$, i.e. $m' + k = (m + k)'$ for every $m \in \mathbb{N}$. Then

$m' + k' = (m' + k)'$ (by property (A)),

$\quad = ((m + k)')'$ (by our supposition that $k \in A$),

$\quad = (m + k')'$ (by property (A)).

This holds for every $m \in \mathbb{N}$, so $k' \in A$. We can therefore apply (P5) to deduce that $A = \mathbb{N}$, i.e., (ii) holds for all m, $n \in \mathbb{N}$.

Now to complete the proof of the theorem we need a further induction. Let B be the set $\{n \in \mathbb{N}: m + n = n + m, \text{for every } m \in \mathbb{N}\}$.

First, $0 \in B$ since $m + 0 = m$ (by property (A)), and $0 + m = m$ (by (i) above). Second, let us suppose that $k \in B$, i.e., $m + k = k + m$ for every $m \in \mathbb{N}$. Now

$$m + k' = (m + k)' \quad \text{(by property (A)),}$$

$$= (k + m)' \quad \text{since } k \in B,$$

$$= k' + m \quad \text{(by (ii) above).}$$

This holds for every $m \in \mathbb{N}$, so $k' \in B$. Applying (P5) to B we conclude that $B = \mathbb{N}$, i.e. $m + n = n + m$ for all $m, n \in \mathbb{N}$.

▶ It is not our purpose to develop elementary number theory, but there are some basic results which we should at least mention.

Remark 1.4

For every natural number n, either $n = 0$ or $n = m'$ for some natural number m.

This may be proved using Peano's axioms. It is left as an exercise for the reader, with the hint that (P5) should be applied to the set $A = \{n \in \mathbb{N} : \text{either } n = 0 \text{ or } n = m' \text{ for some } m \in \mathbb{N}\}$.

Theorem 1.5

Every non-empty set of natural numbers has a least member.

Before we prove this we require to give an explanation of the term 'least member'. Again this is an intuitive notion, but its properties can be derived from the definitions and properties of numbers already given. For $m, n \in \mathbb{N}$ we write $m < n$ if there is $x \in \mathbb{N}$, with $x \neq 0$, such that $m + x = n$. (We also use the notation $m \leq n$, with the obvious meaning.) A set A of natural numbers has a *least member* if there is an element $m \in A$ such that $m < n$ for every other element $n \in A$. The result of Theorem 1.5 is intuitively true when we think of the normal sequence $\{0, 1, 2, \ldots\}$ of natural numbers and note that the relation $<$ corresponds to the relation 'precedes'.

Proof (of Theorem 1.5)

Let A be a set of natural numbers which contains no least member. We show that A is empty. We apply (P5) to the set $B = \{x \in \mathbb{N} : x \leq n \text{ for every } n \in A\}$. Certainly $0 \in B$, since $0 \leq n$ for every $n \in A$. Suppose that $k \in B$. Then $k \leq n$ for every $n \in A$. But k cannot belong to A, since if it did it would be the least element of A. Hence

$k < n$ for every $n \in A$, and consequently $k + 1 \leq n$ for every $n \in A$, i.e. $k + 1 \in B$. By (P5), then, we have $B = \mathbb{N}$. By the definition of B, this means that $x \leq n$ holds for every $x \in \mathbb{N}$ and every $n \in A$. This is impossible unless A is empty, in which case it is vacuously true. The proof is now complete.

▶ The above theorem can be used to justify a slightly different version of the principle of mathematical induction.

(P5*) If A is a set of natural numbers which contains 0 and contains n' whenever $0, 1, \ldots, n$ all belong to A, then A contains all natural numbers.

Theorem 1.6
(P5*) holds (as a consequence of (P5), through Theorem 1.5).

Proof
Let $A \subseteq \mathbb{N}$, with $0 \in A$ and such that $n' \in A$ whenever $0, 1, \ldots$, $n \in A$. We require to show that $A = \mathbb{N}$. Consider the set $\mathbb{N} \backslash A$ (the set of all elements of $\mathbb{N}$ which do not belong to A). Suppose that $\mathbb{N} \backslash A$ is not empty. Then by Theorem 1.5 it contains a least member, n_0, say. We have therefore $n_0 \notin A$, and $x \in A$ for every x with $x < n_0$. Now $n_0 \neq 0$, since $0 \in A$, by our original hypothesis. Hence, $n_0 = m'$ for some $m \in \mathbb{N}$ (by the result of Remark 1.4). So we have $m' \notin A$, but we have also 0, $1, \ldots, m \in A$. This is a contradiction since our hypothesis says that we have $n' \in A$ whenever $0, 1, \ldots, n \in A$. It follows that $\mathbb{N} \backslash A$ is empty, and consequently $A = \mathbb{N}$.

▶ The last of our basic results is one that we shall refer to when we discuss properties of the other number systems. It is the result which is commonly known as the *division algorithm*. Its proof is given here for the sake of completeness, and the reader may omit it.

Theorem 1.7
Let $a \in \mathbb{N}$, $b \in \mathbb{N}$, $b \neq 0$. There exist $q \in \mathbb{N}$, $r \in \mathbb{N}$ with

$$a = qb + r \quad \text{and} \quad r < b.$$

Moreover, the numbers q and r are uniquely determined.

Proof
Let $S = \{y \in \mathbb{N} : y + xb = a, \text{ for some } x \in \mathbb{N}\}$. ($S$ may be thought of as the set of differences $a - xb$ for all those $x \in \mathbb{N}$ such that $a \geq xb$.)

S is not empty, since $a \in S$ (corresponding to $x = 0$). Hence, by Theorem 1.5, S contains a least element, say r. Since $r \in S$, there is $q \in \mathbb{N}$ such that $r + qb = a$, i.e. $a = qb + r$. We must have $r < b$, for otherwise $b \leq r$, so that $r = b + r_1$, say, with $r_1 \in \mathbb{N}$, and $r_1 < r$, necessarily. Then $r + qb = a$ gives $r_1 + b + qb = a$, i.e. $r_1 + (q + 1)b = a$, and this implies that $r_1 \in S$. This contradicts the choice of r as the least element of S.

It remains to show that q and r are unique. Suppose that $a = qb + r = q'b + r'$, with $r < b$, $r' < b$, and $r \leq r'$, say. Then there is $t \in \mathbb{N}$ such that $r + t = r'$, and we have $qb + r = q'b + r + t$, and hence $qb = q'b + t$. It follows that $q'b \leq qb$, so $q' \leq q$. Let $q = q' + u$, say, with $u \in \mathbb{N}$. Then $q'b + ub = qb = q'b + t$, giving $ub = t$. Since $r + t = r'$, then, we have $r + ub = r'$. Consequently $ub \leq r'$, which contradicts $r' < b$, unless $u = 0$. Thus we must have $u = 0$, and this implies that $r = r'$, $t = 0$, and $q = q'$, as required.

In the above proof we have used, besides Theorem 1.5, a few properties of addition, multiplication and inequalities which have not been explicitly derived from our basic assumptions. The most apparent, perhaps, is the cancellation law for inequalities: if $ax \leq bx$ and $x \neq 0$, then $a \leq b$. This may be treated as an exercise.

▶ Natural numbers are a product of intuition. There is no need for a mathematical *definition* of natural numbers. Peano's axioms may be seen as an attempt to define, but they are in fact merely an attempt to characterise natural numbers. But immediately two questions arise. First, are Peano's axioms true of our intuitive natural numbers? And second, is there any collection of objects, essentially different from the set of natural numbers, for which Peano's axioms also hold true? The answer to the first question is clearly (intuitively) in the affirmative. The answer to the second is much harder to find, for it involves the mathematical abstractions: 'collection of objects for which Peano's axioms hold true', and 'essentially different'. We shall see in due course that the second answer is negative, but before that we must explain the abstractions.

Consider the set $2\mathbb{N} = \{2n : n \in \mathbb{N}\}$ of even natural numbers, and denote $k + 2$ by k^*, for each $k \in 2\mathbb{N}$. Then the following are true:

(1) $0 \in 2\mathbb{N}$.

(2) For each $k \in 2\mathbb{N}$, $k^* \in 2\mathbb{N}$.

(3) For no $k \in 2\mathbb{N}$ is k^* equal to 0.

(4) If $k, l \in 2\mathbb{N}$, then $k^* = l^*$.

(5) If $A \subseteq 2\mathbb{N}$ is such that $0 \in A$ and $k^* \in A$ whenever $k \in A$, then $A = 2\mathbb{N}$.

In other words, Peano's axioms 'hold' for the set $2\mathbb{N}$ (together with the operation $*$). It is not difficult to conceive of other structures (i.e. sets together with unary operations) for which Peano's axioms also hold. We can say precisely what this means in general.

> **Definition**
> A *model* of Peano's axioms is a set N, together with a function f and an object e (a triple (N, f, e)) such that
> (P1*) $e \in N$.
> (P2*) The domain of f is N, and for each $x \in N$, $f(x) \in N$.
> (P3*) If $x \in N$, then $f(x) \neq e$.
> (P4*) If $x, y \in N$ and $f(x) = f(y)$, then $x = y$.
> (P5*) If A is a subset of N which contains e and contains $f(x)$ for every $x \in A$, then $A = N$.

The function f is to act like the successor function and e is to act like 0. The reader should compare these conditions (P1*), ..., (P5*) carefully with (P1), ..., (P5).

▶ The model $(2\mathbb{N}, *, 0)$ given above, by its very existence, tells us that Peano's axioms do not characterise the set of natural numbers uniquely. But this new model has a *structure* which is identical to the structure of $(\mathbb{N}, ', 0)$. The two models are *isomorphic*, that is to say there is a bijection $\varphi : \mathbb{N} \to 2\mathbb{N}$ such that $\varphi(n') = (\varphi(n))^*$ for all $n \in \mathbb{N}$, and $\varphi(0) = 0$. (The function φ is given by $\varphi(n) = 2n$.) In general we can make the following definition.

> **Definition**
> Two models (N_1, f_1, e_1) and (N_2, f_2, e_2) of Peano's axioms are *isomorphic* if there is a bijection $\varphi : N_1 \to N_2$ such that
>
> (i) $\varphi(f_1(x)) = f_2(\varphi(x))$,   for all $x \in N_1$,
>
> and
>
> (ii) $\varphi(e_1) = e_2$.

Such a function is said to be an *isomorphism*.

▶ Models of Peano's axioms exist which are different from, but isomorphic to, $(\mathbb{N}, ', 0)$. Mathematically, such models are essentially the same, and for mathematical purposes it really does not matter whether natural numbers are taken to be the elements of $\mathbb{N}$ or the elements of a different but isomorphic model. This will form the basis of our construction of natural numbers within set theory in Section 4.3. In a sense it is only a matter of labelling. If two models are isomorphic then their mathematical characteristics are the same but their elements may be objects of different sorts.

What makes the overall situation sensible, however, is the result of Corollary 1.9 below. It implies that there is no model of Peano's axioms which is not isomorphic to $(\mathbb{N}, ', 0)$. In other words, Peano's axioms do characterise the *structure* of $(\mathbb{N}, ', 0)$ completely.

> **Theorem 1.8** (definition by induction)
> Let (N, f, e) be any model for Peano's axioms. Let X be any set, let $a \in X$ and let g be any function from X to X. Then there is a unique function F from N to X such that
>
> $$F(e) = a,$$
>
> and
>
> $$F(f(x)) = g(F(x)), \quad \text{for each } x \in N.$$

► Theorem 1.8 legitimises what is probably a familiar process for defining functions with domain $\mathbb{N}$. This process was used on page 4 above in the properties (A) and (M). First specify the value of $F(0)$, and then, on the assumption that $F(n)$ has been defined, specify $F(n+1)$ in terms of $F(n)$. Here, of course, we are dealing with an arbitrary model of Peano's axioms, rather than $\mathbb{N}$. The proof of Theorem 1.8 is lengthy and technical, so we shall omit it at this stage. Theorem 4.15 is a particular case of Theorem 1.8, concerning that model of Peano's axioms (the set of abstract natural numbers) which is constructed in Section 4.3. The proof given there can be generalised in a straightforward way to apply to an arbitrary model, as required here.

> **Corollary 1.9**
> Any two models of Peano's axioms are isomorphic.

> *Proof*
> Let (N_1, f_1, e_1) and (N_2, f_2, e_2) be models of Peano's axioms. By Theorem 1.8, there is a unique function $F : N_1 \to N_2$ such that
>
> $$F(e_1) = e_2,$$

and

$$F(f_1(x)) = f_2(F(x)), \quad \text{for each } x \in N_1.$$

This function F thus satisfies conditions (i) and (ii) required by the definition of an isomorphism. It remains only to prove that F is a bijection. Now applying Theorem 1.8 with N_1 and N_2 reversed will yield a unique function $G : N_2 \to N_1$ such that

$$G(e_2) = e_1,$$

and

$$G(f_2(y)) = f_1(G(y)), \quad \text{for each } y \in N_2.$$

We show that $G(F(x)) = x$ for every $x \in N_1$, by application of (P5*) to (N_1, f_1, e_1). Let $A = \{x \in N_1 : G(F(x)) = x\}$. Then $e_1 \in A$, since $G(F(e_1)) = G(e_2) = e_1$. Let $x \in A$. Then $G(F(x)) = x$, so that

$$G(F(f_1(x))) = G(f_2(F(x))) = f_1(G(F(x))) = f_1(x),$$

and consequently $f_1(x) \in A$. It follows, by (P5*), that $A = N_1$. Likewise, we can show that $F(G(y)) = y$ for every $y \in N_2$. Hence, F and G are bijections (and are inverses of each other) and the proof is complete.

▶ The concept of a model of Peano's axioms which is different from, but isomorphic to, $(\mathbb{N}, ', 0)$ is the first stage of mathematical abstraction. Similar abstractions are made in the constructions of the systems of integers, rational numbers, real numbers and complex numbers. These constructions start from the basis of natural numbers and proceed using standard algebraic processes, but in the end they produce sets of mathematical objects which are exceedingly complex in themselves, but which have the necessary properties characterising the number systems in question. What are negative integers? There are some people who argue seriously that they do not exist. But they certainly exist for the mathematician. The mathematician can construct, using his abstract methods, a set which has the properties that the set of integers ought to have, starting from $\mathbb{N}$. We now proceed to do this in some detail. Rationals, reals and complexes will follow.

The way to construct the set of integers is to regard it as the set of all *differences* between ordered pairs of natural numbers. For example:

(2, 3)   gives rise to -1,

(3, 2)   gives rise to 1,

(5, 31)   gives rise to -26, etc.

Notice the significance of the order of the two numbers in the pair. The first problem is that different ordered pairs can give rise to the same integer, for example: $(2, 5)$ and $(7, 10)$ both give rise to -3. Thus we cannot define integers to *be* ordered pairs of natural numbers. What we do is take the collection of all ordered pairs (m, n) with $n - m = 3$ to *represent* the integer -3. We do this via an appropriate equivalence relation, for which the above collection (and all other similarly defined collections) are equivalence classes.

(An equivalence relation on a set X is a binary relation on X which is reflexive, symmetric and transitive. The property which we use is that an equivalence relation gives rise to equivalence classes. An equivalence class consists of all elements of X which are related to a given element. Each element of X determines (and belongs to) one equivalence class. Indeed, X is partitioned into disjoint equivalence classes. We shall mention equivalence relations again in Section 3.1, with more details of the definition. Any standard text on beginning abstract algebra will provide further details if required.)

Now for the formal details of our construction of the integers.

Definition

Let $a, b, c, d \in \mathbb{N}$. We say that (a, b) is related to (c, d), written $(a, b) \Diamond (c, d)$, if $a + d = b + c$. (Notice that we are unable to write '$a - b = c - d$' as we might have wished, because until we have defined negative numbers, differences of natural numbers may not exist.)

Now $\Diamond$ is an equivalence relation. This is easily verified. For any pair (a, b) of natural numbers, $a + b = b + a$, so $(a, b) \Diamond (a, b)$, and $\Diamond$ is reflexive. If $(a, b) \Diamond (c, d)$ then $a + d = b + c$, so $c + b = d + a$, i.e. $(c, d) \Diamond (a, b)$, and $\Diamond$ is symmetric. Lastly, if $(a, b) \Diamond (c, d)$ and $(c, d) \Diamond (e, f)$, then $a + d = b + c$ and $c + f = d + e$. We have

$$a + f + d = a + d + f$$
$$= b + c + f$$
$$= b + d + e$$
$$= b + e + d,$$

and consequently $a + f = b + e$, so that $(a, b) \Diamond (e, f)$, as required to show that $\Diamond$ is transitive.

We define *integers* to be equivalence classes under the relation $\Diamond$. As an example, the set $\{(a, b) : a + 1 = b\}$ is an equivalence class (it is the class determined by $(0, 1)$), and we are *defining* the integer -1 to be this set. However, we shall not use normal notation for integers yet.

Let us denote the equivalence class determined by (a, b) by (a, b). What we intend is that (a, b) should be the integer that we intuitively think of as $a - b$.

▶ All we have so far is a set. It remains to describe the operations of addition and multiplication, to investigate the natural order of the integers and to examine in what way the newly defined set of integers 'contains' the set of natural numbers. This last reflects the way that we normally regard these sets – we do not normally distinguish between natural numbers and non-negative integers.

Definition
Addition and *multiplication* of integers are defined as follows. Let $a, b, c, d \in \mathbb{N}$.

$$(a, b) + (c, d) = (a + c, b + d),$$

$$(a, b) \times (c, d) = (ac + bd, ad + bc).$$

Remarks 1.10
(a) These definitions have an intuitive basis.
$(a - b) + (c - d) = (a + c) - (b + d)$ lies behind the first.
$(a - b) \times (c - d) = (ac + bd) - (ad + bc)$ is the way to remember the second.

(b) We are defining operations on equivalence classes. It is necessary in such a situation to verify that the operations are *well-defined*. We take the case of addition and leave multiplication as an exercise. What we must verify is that if $(a, b) = (p, q)$ and $(c, d) = (r, s)$, then $(a + c, b + d) = (p + r, q + s)$ (i.e. that the result of adding two classes does not depend on the pairs of natural numbers which are chosen to represent them). Suppose that $a + q = b + p$ and $c + s = d + r$. Then

$$(a + c) + (q + s) = (a + q) + (c + s)$$

$$= (b + p) + (d + r)$$

$$= (b + d) + (p + r),$$

and consequently $(a + c, b + d) = (p + r, q + s)$, as required.

(c) If (a, b) is an integer, and $c \in \mathbb{N}$, then $(a + c, b + c) = (a, b)$. To see this, just note that $(a + c, b + c) \diamond (a, b)$, since $(a + c) + b = (b + c) + a$.

(d) It is a straightforward exercise to verify that addition and multiplication satisfy the commutative, associative and distributive laws.

(e) Notice that, for any $a, b \in \mathbb{N}$,

$$(a, b) + (0, 0) = (a, b),$$

$$(a, b) \times (1, 0) = (a, b),$$

and

$$(a, b) \times (0, 0) = (0, 0).$$

Thus $(0, 0)$ behaves like zero, and $(1, 0)$ behaves like 1.

(f) For any $a, b \in \mathbb{N}$,

$$(a, b) + (b, a) = (0, 0).$$

To see this, we need to observe that $(a + b, a + b) = (0, 0)$, which is a special case of the result that $(m, m) = (0, 0)$, for every $m \in \mathbb{N}$.

We write $-(a, b)$ for (b, a), and we abbreviate $(a, b) + (-(c, d))$ by $(a, b) - (c, d)$. Thus we introduce *subtraction* as a legitimate operation on integers.

Exercise

$$(-(a, b)) \times (c, d) = -((a, b) \times (c, d)),$$

$$(-(a, b)) \times (-(c, d)) = (a, b) \times (c, d).$$

Notation

We denote the set of integers by $\mathbb{Z}$, and we shall use variables near the end of the alphabet for elements of $\mathbb{Z}$ (for the time being).

Definition

The order relation on $\mathbb{Z}$ is defined as follows. First we say that an element (a, b) of $\mathbb{Z}$ is *positive* if $b < a$ (as elements of $\mathbb{N}$). Again it must be shown that this is well-defined, i.e. that if $(a, b) = (c, d)$ and $b < a$ then $d < c$. If $a + d = b + c$ and $b < a$ then it certainly follows that $d < c$. $\mathbb{Z}^+$ denotes the set of positive integers. Now we define $<$, for

$x, y \in \mathbb{Z}$ by:

$$x < y \quad \text{if } y - x \in \mathbb{Z}^+.$$

We shall also use the symbol $\leqslant$ (less than or equal to) with its normal meaning.

Remarks 1.11

(a) For $x, y, z \in \mathbb{Z}$ we have $x < y$ if and only if $x + z < y + z$.

(b) For $x, y \in \mathbb{Z}$ and $z \in \mathbb{Z}^+$, we have $x < y$ if and only if $x \times z < y \times z$.

(c) For $x \in \mathbb{Z}$ and $y \in \mathbb{Z}^+$, we have $x < x + y$.

(d) If $x \in \mathbb{Z}^+$ and $y \in \mathbb{Z}^+$, then $x + y \in \mathbb{Z}^+$.

(e) If $x \in \mathbb{Z}^+$ and $y \in \mathbb{Z}^+$, then $x \times y \in \mathbb{Z}^+$.

(f) If $x \in \mathbb{Z}^+$, then $-x < x$.

(g) If $x \in \mathbb{Z}^+$, then $(0, 0) < x$.

(h) For any $x \in \mathbb{Z}$, $(0, 0) \leqslant x^2$.

We sketch proofs for (a) and (e). The others are left as exercises.

For (a), let $x = (a, b)$, $y = (c, d)$, $z = (e, f)$.

$$(y + z) - (x + z) = ((c, d) + (e, f)) - ((a, b) + (e, f))$$

$$= (c + e, d + f) - (a + e, b + f)$$

$$= (c + e, d + f) + (b + f, a + e)$$

$$= (c + e + b + f, d + f + a + e)$$

$$= ((c + b) + (e + f), (d + a) + (e + f))$$

$$= (c + b, d + a)$$

$$= (c, d) + (b, a)$$

$$= (c, d) - (a, b)$$

$$= y - x.$$

Thus $(y + z) - (x + z) \in \mathbb{Z}^+$ if and only if $y - x \in \mathbb{Z}^+$, i.e. $x + z < y + z$ if and only if $x < y$.

For (e), let $x = (a, b)$ and $y = (c, d)$, where $b < a$ and $d < c$.

$$x \times y = (ac + bd, ad + bc).$$

Now there exist $p, q \in \mathbb{N} \backslash \{0\}$ such that $a = b + p$ and $c = d + q$, so

$$ac + bd = (b + p)(d + q) + bd$$

$$= 2bd + pd + bq + pq,$$

and

$$ad + bc = (b+p)d + b(d+q)$$

$$= 2bd + pd + bq.$$

Therefore $ac + bd = ad + bc + pq$, so that

$$ad + bc < ac + bd \text{ in } \mathbb{N} \quad \text{(see Exercise 4 on page 18),}$$

and consequently $x \times y \in \mathbb{Z}^+$.

▶ The pattern that proofs take is well exemplified by the above. Results about elements of $\mathbb{Z}$ are re-stated in terms of equivalence classes of pairs of natural numbers and hence in terms of natural numbers themselves. Properties of $\mathbb{N}$ can then be used to justify properties of $\mathbb{Z}$. Care must be taken in such proofs to distinguish between elements of $\mathbb{Z}$ and elements of $\mathbb{N}$, and to make no assumptions about integers (and, moreover, to avoid treating elements of $\mathbb{N}$ as integers).

The above is a temporary warning only, however. Once the properties of integers have been derived from the properties of natural numbers, we can forget the apparatus of the construction, and treat integers in the intuitive way that we are accustomed to. Part of this intuition is the idea that $\mathbb{N}$ is a subset of $\mathbb{Z}$, i.e. that natural numbers are just non-negative integers. Our construction of $\mathbb{Z}$ renders this convenient idea false. However, we may recover the situation by the following process.

Consider the set S of integers of the form $(n, 0)$ $(n \in \mathbb{N})$. We have seen that $(0, 0)$ behaves like a zero. Let $f : S \to S$ be given by $f(n, 0) = (n+1, 0)$. Then $(S, f, (0, 0))$ is a model for Peano's axioms. This is left for the reader to verify. Moreover, $(S, f, (0, 0))$ is isomorphic to $(\mathbb{N}, ', 0)$, by Corollary 1.9 (the isomorphism associates each $n \in \mathbb{N}$ with $(n, 0) \in S$), and so S has the same mathematical structure as $\mathbb{N}$. Addition and multiplication bear this out, for we know that for $m, n \in \mathbb{N}$,

$$(m, 0) + (n, 0) = (m+n, 0),$$

and

$$(m, 0) \times (n, 0) = (mn, 0).$$

Consequently, we can take the elements of S to represent the natural numbers. This satisfies the formal mathematical requirements. In practice there is no need to do other than just imagine that $\mathbb{N}$ is a subset of $\mathbb{Z}$, in effect regarding n and $(n, 0)$ as different labels for the same object. From now on we actually do so. It should not lead to confusion.

Theorem 1.12

$\mathbb{Z}^+ \cup \{(0,0)\}$, the set of non-negative integers, together with the successor function f given by $f(n,0) = (n+1,0)$ and the zero element $(0,0)$, is a model for Peano's axioms.

Proof

The set S in the above argument is just $\mathbb{Z}^+ \cup \{(0,0)\}$, so the proof is described above. Notice that the non-negative integers thus behave just as natural numbers do.

Theorem 1.13

(i) Given $x \in \mathbb{Z}$, we have *one* of the following: $x \in \mathbb{Z}^+$ or $x = 0$ or $-x \in \mathbb{Z}^+$.

(ii) If $x, y \in \mathbb{Z}^+$, then $x + y \in \mathbb{Z}^+$ and $x \times y \in \mathbb{Z}^+$.

Proof

(i) Let $x = (a,b)$. If $a = b$ then $(a,b) = (0,0)$ and so $x = 0$. Now suppose that $a \neq b$. The set $\{a,b\}$ is a non-empty subset of $\mathbb{N}$, so contains a least member. If the least member is a, then $a < b$. If the least member is b then $b < a$. In the former case we have $-x \in \mathbb{Z}^+$, since $-x = (b,a)$. In the latter case we have $x \in \mathbb{Z}^+$. This proves (i).

(ii) These have already appeared as Remarks 1.11 (d) and (e).

▶ Our purpose here has been to develop the set $\mathbb{Z}$ of integers, and derive its basic properties, from our chosen starting point. This we have now done, and having done so we should forget the apparatus of the construction.

Our procedure merely gives a mathematical way of relating the set of integers to the set of natural numbers, and a demonstration that there is no need to make intuitive assumptions about integers, since our basic assumptions about natural numbers already implicitly contain the standard properties of integers.

With this in mind, from here on integers will be integers and natural numbers will be non-negative integers. The next stage in our development is a very similar construction, the construction of the set of rational numbers.

Exercises

1. Using (P5), prove the following: if $P(n)$ is a statement about the natural number n such that $P(0)$ holds and $P(n')$ holds whenever $P(n)$ holds, then $P(n)$ holds for every natural number n.

2. Verify that addition on $\mathbb{N}$ is associative.
3. Verify that multiplication on $\mathbb{N}$ is commutative and associative and that the usual distributive law holds.
4. Prove that for every natural number n, either $n = 0$ or $n = m'$ for some natural number m. Hence, show that the product of two non-zero natural numbers is non-zero.
5. Prove that for every pair of natural numbers m and n, either $m \leq n$ or $n \leq m$.
6. Let $m, n \in \mathbb{N}$, with $m \neq 0$. Prove that there exists $r \in \mathbb{N}$ such that $n < rm$. (Hint: use Theorem 1.7.)
7. Show that multiplication of integers is well-defined, i.e. that if $(a, b) = (p, q)$ and $(c, d) = (r, s)$ then $(ac + bd, ad + bc) = (pr + qs, ps + qr)$.
8. Verify the commutative, associative and distributive laws for addition and multiplication on $\mathbb{Z}$.
9. Prove Remarks 1.11(b), (c), (d), (f), (g) and (h).
10. Let a be a fixed element of $\mathbb{Z}$. Let A be a subset of $\mathbb{Z}$ such that $a \in A$ and $x + 1 \in A$ whenever $x \in A$. Prove that $\{x \in \mathbb{Z} : a \leq x\} \subseteq A$.
11. Prove that every non-empty set of integers which is bounded below has a least element.
12. Prove that every non-empty set of integers which is bounded above has a greatest element.
13. Prove that for any pair of integers a and b, either $a \leq b$ or $b \leq a$.
14. Let $x, y \in \mathbb{Z}$ be such that $xy = 0$. Prove that $x = 0$ or $y = 0$.

1.2 Rational numbers

There are four standard arithmetic operations: addition, sub-traction, multiplication and division. In $\mathbb{N}$ only the first and third are permitted in general, since it need not be the case, for natural numbers a and b, that $a - b$ or a/b are natural numbers. The set $\mathbb{Z}$ of integers is such that subtraction is permitted, but it is still the case that division may not work in $\mathbb{Z}$. Just as we took differences of natural numbers to represent integers, here the essence of the process is to use ordered pairs representing quotients. The standard way of representing rational numbers is as quotients of integers. Of course, the same rational number may be represented thus in many different ways. Consequently, in our formal procedure, the pairs $(2, 3)$, $(8, 12)$, $(-50, -75)$ and $(1000, 1500)$ will all represent the same object. This makes sense intuitively if we think of them as representing the familiar object $2/3$. The formal details are similar to those of the earlier construction of the integers.

Definition

Let $a, c \in \mathbb{Z}$, and let $b, d \in \mathbb{Z}\backslash\{0\}$. We say that (a, b) is related to (c, d), written $(a, b) \# (c, d)$, if $ad = bc$. (Notice that this expresses what we would like, namely $a/b = c/d$, but as yet we cannot write fractions since we do not have a division operation on $\mathbb{Z}$.) Intuitively we have $(a, b) \mathrel{\#} (c, d)$ if a/b and c/d represent the same rational number.

Now $\#$ is an equivalence relation. First, for any $a, b \in \mathbb{Z}$ with $b \neq 0$, we have $ab = ab$, so $(a, b) \mathrel{\#} (a, b)$, and so $\#$ is reflexive. Second, suppose that $(a, b) \mathrel{\#} (c, d)$, so that $ad = bc$. Then $cb = da$ clearly, so $(c, d) \mathrel{\#} (a, b)$, and we have shown that $\#$ is symmetric. Third, suppose that $(a, b) \mathrel{\#} (c, d)$ and $(c, d) \mathrel{\#} (e, f)$, where $a, c, e \in \mathbb{Z}$ and $b, d, f \in \mathbb{Z} \backslash \{0\}$. Then $ad = bc$ and $cf = de$. We have

$$afd = adf = bcf = bde = bed,$$

so since $d \neq 0$ we can deduce $af = be$. Hence, $(a, b) \mathrel{\#} (e, f)$ as required to show that $\#$ is transitive.

We define the set of *rational numbers* to be the set of equivalence classes under $\#$. As an example, the set $\{(a, b) : a, b \in \mathbb{Z}, b \neq 0, b = 2a\}$ is an equivalence class (it is the class determined by $(1, 2)$).

Let us denote the equivalence class determined by (a, b) by a/b. What we intend is that a/b should be the rational number that we intuitively think of as a/b.

▶ Our exposition has been deliberately modelled on the previous description of the construction of the integers, so as to emphasise the analogy. In algebraic terms, the construction of the integers involved introducing 'additive inverses' for the natural numbers (namely, negative integers), and now the construction of the rational numbers involves the introduction of 'multiplicative inverses' for the non-zero integers. In this case, of course, we must also introduce other new objects; besides requiring rational numbers of the form $1/b (b \in \mathbb{Z}, b \neq 0)$ we also have rationals of the form a/b which cannot be reduced (by cancellation) to a fraction with numerator 1.

Again, all we have so far is a set. It remains to describe the operations of addition and multiplication (and subtraction and division), to investigate the natural order of the rational numbers, and to examine the way in which the newly-defined set of rational numbers contains the set of integers.

Definition

Addition and *multiplication* of rational numbers are defined as follows. Let $a, b, c, d \in \mathbb{Z}$, with $b \neq 0, d \neq 0$.

$$a/b + c/d = (ad + bc)/bd,$$

$$a/b \times c/d = ac/bd.$$

Remarks 1.14

(a) The above definitions reflect our intuitive basis for rational numbers. We think of

$$\frac{a}{b} + \frac{c}{d} = \frac{ad+bc}{bd}, \quad \text{and} \quad \frac{a}{b} \times \frac{c}{d} = \frac{ac}{bd}.$$

(b) We must verify that these operations are well-defined. This time we take the case of multiplication and leave addition as an exercise. Suppose that $a/b = p/q$ and $c/d = r/s$. We must show that $ac/bd = pr/qs$, i.e. that $(ac, bd) \# (pr, qs)$, i.e. that $acqs = bdpr$. Now we have supposed that $a/b = p/q$, and consequently that $aq = bp$, and similarly we have $cs = dr$. Thus

$$acqs = aqcs = bpdr = bdpr,$$

as required.

(c) If a/b is a rational number, and x is a non-zero integer, then $ax/bx = a/b$. To see this we just note that $(ax, bx) \# (a, b)$, since $axb = bxa$.

(d) Addition and multiplication of rational numbers are commutative and associative, and the distributive law holds. These results are easy consequences of the corresponding properties of integers. To illustrate, let us take the distributive law. Let $a, b, c, d, e, f \in \mathbb{Z}$, with $b \neq 0$, $d \neq 0$, $f \neq 0$.

$$(a/b) \times (c/d + e/f)$$

$$= (a/b) \times ((cf + de)/df)$$

$$= a(cf + de)/bdf$$

$$= (acf + ade)/bdf.$$

Also

$$(a/b \times c/d) + (a/b \times e/f)$$

$$= (ac/bd) + (ae/bf)$$

$$= (acbf + bdae)/bdbf$$

$$= b(acf + ade)/b(bdf)$$

$$= (acf + ade)/bdf, \text{ by (c) above, since } b \neq 0,$$

$$= (a/b) \times (c/d + e/f) \text{ from above.}$$

(e) We have rational numbers which behave like zero and one. For any $a, b \in \mathbb{Z}$ with $b \neq 0$, we have:

$$a/b + 0/1 = a/b,$$

$$a/b \times 1/1 = a/b,$$

$$a/b \times 0/1 = 0/1.$$

(The last requires a one-step proof.)
Thus $0/1$ behaves like zero. Notice that $0/b$ is equal to $0/1$, for any non-zero integer b. Also $1/1$ behaves like 1, and for any non-zero integer b we have $b/b = 1/1$.

(f) For any $a, b \in \mathbb{Z} \backslash \{0\}$, we have

$$a/b \times b/a = ab/ab = 1/1.$$

Thus b/a is a multiplicative inverse of a/b, and this enables us to introduce the operation of *division*. Division by a/b is defined to be the same as multiplying by b/a. Note, of course that we can do this only if b/a is a rational number, i.e. only if $a \neq 0$. Hence, the restriction (required by intuition, of course) that we can divide only by non-zero numbers. We shall later use the normal notation for division and for fractions, but expressions like

$$\frac{1}{a/b}, \quad \text{and} \quad \frac{c/d}{a/b}$$

are rather cumbersome, and we shall try to avoid them.

(g) Additive inverses are straightforward. For $a, b \in \mathbb{Z}$ with $b \neq 0$ we have

$$a/b + (-a)/b = (ab - ab)/b^2 = 0/b^2 = 0/1.$$

Hence, we may write $-(a/b)$ for $(-a)/b$. Observe that $(-a)/b = a/(-b)$, which of course fits with our intuitive ideas.

(h) *Subtraction.* For $a, b, c, d \in \mathbb{Z}$, with $b \neq 0$, $d \neq 0$, let $a/b - c/d$ stand for $a/b + (-(c/d))$.

▶ We denote the set of rational numbers by $\mathbb{Q}$. As before, we adopt a temporary convention that letters near the beginning of the alphabet denote integers and letters near the end will denote rational numbers.

Definition
The order relation on $\mathbb{Q}$ is defined as follows. We say that an element a/b of $\mathbb{Q}$ is *positive* if $ab > 0$ (ab is of course an integer). This is well-defined, for suppose that $a/b = c/d$ and a/b is positive. Then $ad = bc$ and $ab > 0$. It follows that

$$cdb^2 = bcbd = adbd = abd^2.$$

Since $b^2 > 0$, $d^2 > 0$ and $ab > 0$, we must have $cd > 0$.

The set of positive rational numbers is denoted by $\mathbb{Q}^+$. Now we define $<$ on $\mathbb{Q}$ by:

$$x < y \text{ if } y - x \in \mathbb{Q}^+.$$

The definition of $\leqslant$ is now just as one would expect (less than or equals).

Remarks 1.15 (see Remarks 1.11)
(a) For $x, y, z \in \mathbb{Q}$, we have $x < y$ if and only if $x + z < y + z$.
(b) For $x, y \in \mathbb{Q}$ and $z \in \mathbb{Q}^+$, we have $x < y$ if and only if $x \times z < y \times z$.
(c) For $x \in \mathbb{Q}$ and $y \in \mathbb{Q}^+$, we have $x < x + y$.
(d) If $x \in \mathbb{Q}^+$ and $y \in \mathbb{Q}^+$, then $x + y \in \mathbb{Q}^+$.
(e) If $x \in \mathbb{Q}^+$ and $y \in \mathbb{Q}^+$, then $x \times y \in \mathbb{Q}^+$.
(f) If $x \in \mathbb{Q}^+$, then $-x < x$.
(g) $x \in \mathbb{Q}^+$ if and only if $0/1 < x$.
(h) For any $x \in \mathbb{Q}$, $0/1 \leqslant x^2$.

We sketch proofs for (b) and (d). The others are left as exercises.

For (b), let $x = a/b$, $y = c/d$, $z = e/f$, where $b \neq 0$, $d \neq 0$, $f \neq 0$, and $ef > 0$.

$$y - x = c/d + (-a)/b$$

$$= (cb - da)/db,$$

so $x < y$ if and only if $(cb - da)db > 0$. Also,

$$y \times z - x \times z = ce/df + (-ae)/bf$$

$$= (cebf - dfae)/dfbf,$$

so $x \times z < y \times z$ if and only if $(cebf - dfae)dfbf > 0$. Now

$$(cebf - dfae)dfbf = (cb - da)bdef^3$$

$$= (cb - da)bd(ef)f^2.$$

We know that $ef>0$ and $f^2>0$, so the left-hand side is positive (as an integer) if and only if $(cb-da)bd>0$. The result follows.

For (d), let $x=a/b$, and $y=c/d$, where $b\neq0, d\neq0$, $ab>0$, and $cd>0$.

$$x+y=a/b+c/d$$

$$=(ad+bc)/bd.$$

We require, therefore, $(ad+bc)bd>0$. Now $(ad+bc)bd=abd^2+cdb^2$, and by our supposition $ab>0$ and $cd>0$. Since $d^2>0$ and $b^2>0$, the result follows.

▶ The reader who is familiar with a little abstract algebra will know that the set of rational numbers, with the operations of addition and multiplication, is a field. We have not emphasised particular algebraic properties, so it would be a useful exercise to verify that our set $\mathbb{Q}$, with the operations that we have defined, does satisfy the requirements for a field. It is also the case that our set $\mathbb{Z}$, with its addition and multiplication operations, is an integral domain. The property that a field has, which an integral domain does not necessarily have, is that every non-zero element has a multiplicative universe. Our construction of $\mathbb{Q}$ was designed to meet this requirement and, indeed, the construction may be carried through with only trivial modifications, starting with an arbitrary integral domain D in place of $\mathbb{Z}$, and finishing with a field in which D can be embedded. We shall not pursue this, but again it would be a useful exercise for the reader with some knowledge of algebra.

Let us return to our specific construction.

It is the case that $\mathbb{Z}$ can be *embedded* in $\mathbb{Q}$. We saw this situation before with regard to $\mathbb{N}$ and $\mathbb{Z}$, and we shall adopt the same convention here, in order to regard $\mathbb{Z}$ as actually a subset of $\mathbb{Q}$. First, which elements of $\mathbb{Q}$ are the ones which behave like integers? We have already noted $0/1$ and $1/1$. It is not hard to guess that elements of the form $a/1$ (with $a\in\mathbb{Z}$) will constitute a 'copy' of $\mathbb{Z}$ inside $\mathbb{Q}$. Let us check the operations.

$$a/1+b/1=(a+b)/1,$$

and

$$a/1\times b/1=ab/1,$$

so these elements of $\mathbb{Q}$ do indeed behave as though they were integers. Also, there is a clear correspondence between integers a and rational numbers $a/1$.

There is in fact an isomorphism between $\mathbb{Z}$ and the subsystem $\{a/1 : a \in \mathbb{Z}\}$ of $\mathbb{Q}$. This subset of $\mathbb{Q}$ therefore has the same mathematical structure as $\mathbb{Z}$. Just as with $\mathbb{N}$ and $\mathbb{Z}$ earlier we can regard $\mathbb{Z}$ as a subset of $\mathbb{Q}$ by taking the set $\{a/1 : a \in \mathbb{Z}\}$ as a representation of the system of integers, and in practice forgetting the distinction between elements a of $\mathbb{Z}$ and elements $a/1$ of $\mathbb{Q}$.

$\mathbb{Q}$ is not merely a field. It has a natural ordering of its elements which has convenient properties in relation to the operations of addition and multiplication. Some of these are best expressed in terms of the set $\mathbb{Q}^+$ rather than the order $<$ itself (recall, of course, that the definition of $\mathbb{Q}^+$ was an essential part of the definition of $<$).

Theorem 1.16

(i) Given $x \in \mathbb{Q}$, *one* of the following holds:

$$x \in \mathbb{Q}^+, x = 0, -x \in \mathbb{Q}^+.$$

(ii) If $x, y \in \mathbb{Q}^+$, then $x + y \in \mathbb{Q}^+$ and $x \times y \in \mathbb{Q}^+$.

Proof

(i) Let $x = a/b$, with $a, b \in \mathbb{Z}$, $b \neq 0$. If $a = 0$ then $x = 0/b$, i.e. $x = 0$ as a rational number. If $a \neq 0$ then we apply Theorem 1.13 to a and b and examine the possibilities separately. If $a \in \mathbb{Z}^+$ and $b \in \mathbb{Z}^+$ then $ab \in \mathbb{Z}^+$, so $x \in \mathbb{Q}^+$. If $a \in \mathbb{Z}^+$ and $-b \in \mathbb{Z}^+$ then $(-ab) \in \mathbb{Z}^+$, so $-x \in \mathbb{Q}^+$. The other two cases are equally straightforward. The reader is left to complete the proof.

(ii) These have appeared before as Remarks 1.15 (d) and (e).

▶ There are two other basic properties of $\mathbb{Q}$, which we should mention before proceeding to discuss the set of real numbers. These are the density property and the Archimedean property.

Theorem 1.17

The natural ordering of $\mathbb{Q}$ is *dense*, i.e. given $x, y \in \mathbb{Q}$ with $x < y$, there is $z \in \mathbb{Q}$ such that $x < z$ and $z < y$.

Proof

Let $x, y \in \mathbb{Q}$ with $x < y$. Take $z = (x + y)/2$. Then

$$z - x = \frac{y - x}{2} \in \mathbb{Q}^+ \quad \text{since } y - x \in \mathbb{Q}^+.$$

Also

$$y - z = \frac{y - x}{2} \in \mathbb{Q}^+.$$

We have, therefore, $x < z$ and $z < y$.

Theorem 1.18
The natural ordering of $\mathbb{Q}$ is *Archimedean*, i.e. given $x, y \in \mathbb{Q}^+$, there is a positive integer n such that $y < nx$.

Proof
Let $x = a/b$, $y = c/d$, with $a, b, c, d \in \mathbb{Z}$ and $b \neq 0$, $d \neq 0$. Now $x, y \in \mathbb{Q}^+$ so we may assume that $a, b, c, d \in \mathbb{Z}^+$. This is because $ab > 0$ and $cd > 0$ imply that a and b have the same sign, and so do c and d. If they are not already positive, we can take x to be $(-a)/(-b)$ or y to be $(-c)/(-d)$ or both. Let us now recall what we are seeking. We require $n/1 \in \mathbb{Q}$ such that

$$c/d < (n/1) \times (a/b),$$

i.e.

$$c/d < na/b,$$

i.e.

$$na/b - c/d \in \mathbb{Q}^+,$$

i.e.

$$(nad - bc)/bd \in \mathbb{Q}^+,$$

and this will be the case, provided that $(nad - bc)bd$ is a positive integer. But $b \in \mathbb{Z}^+$ and $d \in \mathbb{Z}^+$, so we require to find $n \in \mathbb{Z}^+$ such that $nad - bc \in \mathbb{Z}^+$. By Theorem 1.13 (ii), $ad \in \mathbb{Z}^+$ and $bc \in \mathbb{Z}^+$. By Theorem 1.7, there exist $q \in \mathbb{N}$ and $r \in \mathbb{N}$ with $bc = q(ad) + r$, and $0 \leqslant r < ad$. Consequently,

$$bc < q(ad) + ad,$$

i.e.

$$bc < (q + 1)ad.$$

Now take n to be $q + 1$, so that we have

$$nad - bc \in \mathbb{Z}^+,$$

as required. This completes the proof.

Exercises

1. Verify that the addition operation on $\mathbb{Q}$ is well-defined.
2. Prove that addition and multiplication are both commutative and associative operations on $\mathbb{Q}$.
3. Prove Remarks 1.15 (a), (c), (e), (f), (g) and (h).
4. Let $x \in \mathbb{Q}^+$. Show that x may be written as a/b, with $a, b \in \mathbb{Z}^+$.
5. Prove that for every pair of rational numbers x and y, either $x \le y$ or $y \le x$.
6. Let $x \in \mathbb{Q}$. Prove that there is an integer which is greater than x. Prove also that there is an integer which is smaller than x.
7. Let $x, y \in \mathbb{Q}$ be such that $xy = 0$. Prove that $x = 0$ or $y = 0$.
8. Outline a construction which yields, given any integral domain D, a field F and an embedding of D in F. (Hint: $\mathbb{Z}$ is an integral domain and $\mathbb{Q}$ is a field in which $\mathbb{Z}$ is embedded, so just follow the same process. This exercise, of course, cannot be attempted by those unfamiliar with abstract algebraic notions involved.)

1.3 Real numbers

The set of rational numbers suffers from some mathematical limitations. This has been known at least since Pythagoras' time, as it is he who is generally credited with the first proof that there is no rational number whose square is 2. This presents a difficulty in geometry, for example, for it means that a square with sides of length 1 unit has a diagonal whose length is not a rational number. How then is such a length to be regarded? How, indeed, is it to be specified or represented? The first of these questions is difficult. Mathematicians got by for centuries without really addressing themselves to it, while at the same time using the convenient specification 'the number whose square is 2' and the convenient representation $\sqrt{2}$.

This is a particular case of a more general limitation of $\mathbb{Q}$. Polynomial equations with integer coefficients may not have solutions in $\mathbb{Q}$. In this case the equation is $x^2 - 2 = 0$. Of course, the equation $ax^2 + bx + c = 0$ does not have a solution in $\mathbb{Q}$ unless $b^2 - 4ac$ is a perfect square ($a, b, c \in \mathbb{Z}$).

But the irrationality of $\sqrt{2}$ is also a particular case of another general limitation of $\mathbb{Q}$. This is that convergent sequences may not have limits. Another way of expressing this is that subsets of $\mathbb{Q}$, which are bounded above, need not have least upper bounds. Let us illustrate these ideas, again using $\sqrt{2}$.

The sequence

$$1, 1.4, 1.41, 1.414, 1.414\,2, 1.414\,21, \ldots,$$

obtained by truncating the (infinite) decimal representation of $\sqrt{2}$, is a sequence of rational numbers. It is convergent in the sense that the differences between the terms of the sequence approach zero. Nevertheless it has no limit in $\mathbb{Q}$, for the limit would have to be $\sqrt{2}$.

The set $\{x \in \mathbb{Q} : x^2 < 2\}$ is a subset of $\mathbb{Q}$ which is bounded above (i.e. there is a rational number greater than every member of the set). It has no least upper bound in $\mathbb{Q}$. This is not easy to verify and, for the moment, we leave it on an intuitive level.

The construction of the set of real numbers is based on the same algebraic principles as the constructions of the other number systems, in that we define an equivalence relation on a particular set and take the equivalence classes to be our 'new' numbers. However, the situation here is more complicated because this time it is not the provision of additive or multiplicative inverses which is the purpose of the construction. Here we must ensure that each non-empty set of real numbers, which is bounded above, has a least upper bound. We must, in effect, insert a least upper bound for each such set. The construction itself, however, tends to obscure this purpose, since, as in the other constructions, we produce an entirely new set of objects and go on to see that the original set (in this case $\mathbb{Q}$) can be embedded in the new set in a natural way.

Let us consider sequences again. In our example above, we had the sequence

$$1, 1.4, 1.41, 1.414, 1.4142, \ldots,$$

which 'converges' to $\sqrt{2}$. In a similar way,

$$2, 1.5, 1.42, 1.415, 1.4143, \ldots$$

'converges' to $\sqrt{2}$, but from above. Certainly there are many different sequences with this same property. What we shall do is take $\sqrt{2}$ to *be* the collection of all sequences of rational numbers which converge in this sense to $\sqrt{2}$. This requires to be made more precise, in order to avoid circularity.

Definition

A sequence $x_1, x_2, x_3, \ldots$ of rational numbers is a *Cauchy sequence* if $x_m - x_n \to 0$ as $m, n \to \infty$. More precisely, this says: given any positive rational number ε, there exists a positive integer N such that

$$|x_m - x_n| < \varepsilon, \text{ for all } m, n > N.$$

► This makes precise our usage above of the word 'convergent'. The sequence 1, 1.4, 1.41, . . . of approximations to $\sqrt{2}$ is a Cauchy sequence. The difference between the term with m digits after the decimal point and the term with n digits after the decimal point (say $n > m$) is less than 1 unit in the mth digit, i.e. less than $1/10^m$. So, given any $\varepsilon \in \mathbb{Q}^+$, choose N so that $1/10^N < \varepsilon$, and thus $|x_m - x_n| < \varepsilon$, for all $m, n > N$.

This same demonstration will apply similarly to any such sequence of decimal approximations. For example, by calculating π with increasing accuracy, we may obtain the sequence 3, 3.1, 3.14, 3.141, 3.1415, . . . , which is a Cauchy sequence with no limit in $\mathbb{Q}$.

There is no reason to restrict ourselves here to Cauchy sequences without limits. It is easily seen that 0, 0.3, 0.33, 0.333, 0.3333, . . . is a Cauchy sequence, and it has a limit in $\mathbb{Q}$, namely, $1/3$. So it will be not only irrational numbers which are represented by a corresponding collection of Cauchy sequences. All rational numbers will also be so represented, and this will give us our embedding of $\mathbb{Q}$ in $\mathbb{R}$.

In order to give our definition of real numbers, it remains to make precise, without circularity, what it means for two Cauchy sequences to have the same limit, and so represent the same real number. The point to note is that we cannot mention 'limit' in general for Cauchy sequences in $\mathbb{Q}$, since limits do not always exist in $\mathbb{Q}$. We can get around this, however.

Definition

Let us denote sequences $a_1, a_2, \ldots$ and $b_1, b_2, \ldots$ by (a_n) and (b_n). We say that two Cauchy sequences (a_n) and (b_n) are *equivalent*, and write $(a_n) \approx (b_n)$, if the sequence whose nth term is $a_n - b_n$ converges to zero.

Recall that a sequence (x_n) converges to zero if, given any $\varepsilon \in \mathbb{Q}^+$, there exists $N \in \mathbb{N}$ such that $|x_n| < \varepsilon$ for all $n > N$.

Example 1.19

Take the sequences

$$1, 1.4, 1.41, 1.414, 1.414\,2, 1.414\,21, \ldots,$$

$$2, 1.5, 1.42, 1.415, 1.414\,3, 1.414\,22, \ldots.$$

The sequence of differences (second minus first) is

$$1, 0.1, 0.01, 0.001, 0.000\,1, 0.000\,01, \ldots.$$

The nth term of this sequence is $1/10^{n-1}$, and the sequence clearly converges to zero.

Theorem 1.20

The relation $\approx$ is an equivalence relation on the set of all Cauchy sequences in $\mathbb{Q}$.

Proof

$\approx$ is trivially reflexive and symmetric. Let us demonstrate transitivity. Suppose that (a_n), (b_n) and (c_n) are Cauchy sequences, and that $(a_n) \approx (b_n)$ and $(b_n) \approx (c_n)$. Then, if we set $x_n = a_n - b_n$ and $y_n = b_n - c_n$, we know that (x_n) and (y_n) both converge to zero. Consequently, the sequence whose nth term is $x_n + y_n$ converges to zero. But

$$x_n + y_n = a_n - b_n + b_n - c_n = a_n - c_n.$$

Thus $(a_n) \approx (c_n)$, as required.

Definition

The set $\mathbb{R}$ of *real numbers* is the set of all the equivalence classes of Cauchy sequences in $\mathbb{Q}$ under the relation $\approx$.

▶ So far so good. Most of the work lies ahead of us, however, since we have still to verify that this set has the properties that we expect the set of real numbers to have. The principal properties concern addition and multiplication, the natural order of the real numbers, and the way in which $\mathbb{Q}$ is embedded in $\mathbb{R}$.

Notation

We denote by $[a_n]$ the equivalence class containing the Cauchy sequence (a_n).

Definitions

Addition and *multiplication* of real numbers are defined by

$$[x_n] + [y_n] = [x_n + y_n],$$

and

$$[x_n] \times [y_n] = [x_n y_n],$$

where (x_n) and (y_n) are Cauchy sequences in $\mathbb{Q}$, and $(x_n + y_n)$ and $(x_n y_n)$ denote the sequences whose nth terms are respectively $x_n + y_n$ and $x_n y_n$.

▶ There is much to be verified before these definitions can be accepted. We list the necessary results in a theorem.

Theorem 1.21

(i) Let (x_n) and (y_n) be Cauchy sequences in $\mathbb{Q}$. Then $(x_n + y_n)$ and $(x_n y_n)$ are also Cauchy sequences in $\mathbb{Q}$.

(ii) Let (x_n), (y_n), (x'_n), (y'_n) be Cauchy sequences in $\mathbb{Q}$ with $(x_n) \approx (x'_n)$ and $(y_n) \approx (y'_n)$. Then

$$(x_n + y_n) \approx (x'_n + y'_n), \text{ and } (x_n y_n) \approx (x'_n y'_n).$$

Proof

(i) Let (x_n) and (y_n) be Cauchy sequences in $\mathbb{Q}$. Choose a positive rational number ε. There is $N_1 \in \mathbb{N}$ such that $|x_m - x_n| < \frac{1}{2}\varepsilon$ for all $m, n > N_1$. Also, there is $N_2 \in \mathbb{N}$ such that $|y_m - y_n| < \frac{1}{2}\varepsilon$ for all $m, n > N_2$. Hence, for all $m, n > \max(N_1, N_2)$, we have

$$|(x_m + y_m) - (x_n + y_n)| = |(x_m - x_n) + (y_m - y_n)|$$

$$\leq |x_m - x_n| + |y_m - y_n|$$

$$< \tfrac{1}{2}\varepsilon + \tfrac{1}{2}\varepsilon = \varepsilon.$$

Thus $(x_n + y_n)$ is a Cauchy sequence, as required.

The proof that $(x_n y_n)$ is a Cauchy sequence uses the result of the lemma which follows this proof. By the lemma there exist rational numbers d and e such that

$$|x_n| \leq d \quad \text{and} \quad |y_n| \leq e, \quad \text{for all } n \in \mathbb{N}.$$

Choose a positive rational number ε. There exists $N_1 \in \mathbb{N}$ such that

$$|x_m - x_n| < \frac{\varepsilon}{2e}, \quad \text{for all } m, n > N_1.$$

Also, there exists $N_2 \in \mathbb{N}$ such that

$$|y_m - y_n| < \frac{\varepsilon}{2d}, \quad \text{for all } m, n > N_2.$$

Hence, for all $m, n > \max(N_1, N_2)$, we have

$$|x_m y_m - x_n y_n| = |x_m y_m - x_n y_m + x_n y_m - x_n y_n|$$

$$= |(x_m - x_n)y_m + x_n(y_m - y_n)|$$

$$\leq |(x_m - x_n)y_m| + |x_n(y_m - y_n)|$$

$$= |x_m - x_n||y_m| + |x_n||y_m - y_n|$$

$$< \frac{\varepsilon}{2e} \times e + d \times \frac{\varepsilon}{2d} = \varepsilon.$$

Thus $(x_n y_n)$ is a Cauchy sequence, as required.

(ii) The case of addition is left as an exercise for the reader. Suppose that (x_n), (y_n), (x_n'), (y_n') are Cauchy sequences, and that $(x_n) \approx (x_n')$ and $(y_n) \approx (y_n')$. We follow an argument similar to the preceding one. Choose a positive rational number ε. The sequences (x_n) and (y_n') are bounded, say

$$|x_n| \le k \quad \text{and} \quad |y_n'| \le l, \quad \text{for all } n \in \mathbb{N}.$$

There exists $N_1 \in \mathbb{N}$ such that

$$|x_n - x_n'| < \frac{\varepsilon}{2l}, \quad \text{for all } n > N_1,$$

since $(x_n) \approx (x_n')$. Similarly, there exists $N_2 \in \mathbb{N}$ such that

$$|y_n - y_n'| < \frac{\varepsilon}{2k}, \quad \text{for all } n > N_2.$$

Hence, for all $n > \max(N_1, N_2)$, we have

$$|x_n y_n - x_n' y_n'| = |x_n y_n - x_n y_n' + x_n y_n' - x_n' y_n'|$$

$$= |x_n(y_n - y_n') + (x_n - x_n')y_n'|$$

$$\le |x_n(y_n - y_n')| + |(x_n - x_n')y_n'|$$

$$= |x_n||y_n - y_n'| + |x_n - x_n'||y_n'|$$

$$< k \times \frac{\varepsilon}{2k} + \frac{\varepsilon}{2l} \times l = \varepsilon.$$

Thus $(x_n y_n) \approx (x_n' y_n')$, as required.

Lemma
Every Cauchy sequence in $\mathbb{Q}$ is bounded, i.e. given a Cauchy sequence (a_n) there is a rational number d such that $|a_n| \le d$ for every $n \in \mathbb{N}$.

Proof
Let (a_n) be a Cauchy sequence, and choose any positive rational number ε. Then there is $N \in \mathbb{N}$ such that $|a_m - a_n| < \varepsilon$ for all $m, n > N$. It follows that

$$|a_n| < |a_{N+1}| + \varepsilon, \quad \text{for all } n > N.$$

We thus have a bound on all terms in the sequence after the Nth. It may happen that one of the earlier terms is larger than this bound, so let $d = \max(|a_1|, |a_2|, \ldots, |a_N|, |a_{N+1}| + \varepsilon)$. Then $|a_n| \leq d$ for all $n \in \mathbb{N}$ as required.

▶ The definitions of addition and multiplication on $\mathbb{R}$ now have been shown to make sense. Next we list some elementary properties.

Theorem 1.22

(i) Addition and multiplication on $\mathbb{R}$ are commutative and associative, and they satisfy the distributive law.

(ii) The equivalence class of Cauchy sequences which contains the constant sequence $0, 0, 0, \ldots$ behaves additively and multiplicatively like zero. We denote this by $[0]$.

(iii) The equivalence class of Cauchy sequences which contains the constant sequence $1, 1, 1, \ldots$ behaves multiplicatively like a 1. We denote this by $[1]$.

(iv) If $[a_n] \in \mathbb{R}$, then $[a_n] + [-a_n] = [0]$.

(v) If $[a_n] \in \mathbb{R}$ and $[a_n] \neq [0]$, then there is $[b_n] \in \mathbb{R}$ such that $[a_n] \times [b_n] = [1]$. Note that we cannot just set $b_n = 1/a_n$, since we could have $a_n = 0$ for some values of n.

(These properties may be summarised by the assertion that $\mathbb{R}$ is a field, under the given addition and multiplication operations.)

Proof

(i) We prove one of these results and leave the others as exercises. Let $[a_n], [b_n], [c_n] \in \mathbb{R}$.

$$[a_n] \times ([b_n] + [c_n])$$

$$= [a_n] \times [b_n + c_n]$$

$$= [a_n(b_n + c_n)]$$

$$= [a_n b_n + a_n c_n], \quad \text{by the distributive law in } \mathbb{Q},$$

$$= [a_n] \times [b_n] + [a_n] \times [c_n], \quad \text{as required.}$$

(ii) Let $[a_n] \in \mathbb{R}$.

$$[a_n] + [0] = [a_n + 0] = [a_n],$$

and

$$[a_n] \times [0] = [a_n \times 0] = [0].$$

(iii) Let $[a_n] \in \mathbb{R}$.

$$[a_n] \times [1] = [a_n \times 1] = [a_n].$$

(iv) Trivial. Note that this enables us to define subtraction on $\mathbb{R}$.

$[a_n] - [b_n]$ means $[a_n] + [-b_n]$.

(v) This requires rather more effort. Let $[a_n] \in \mathbb{R}$, with $[a_n] \neq [0]$. Then (a_n) does not have limit zero, so by the lemma which follows this proof, there exist $e \in \mathbb{Q}^+$ and $K \in \mathbb{N}$ such that

$$|a_n| \geqslant e \quad \text{for all } n > K.$$

We define our sequence (b_n) by:

$$b_n = \begin{cases} 0 & \text{for } n \leqslant K \\ \dfrac{1}{a_n} & \text{for } n > K. \end{cases}$$

Now we must show that (b_n) is a Cauchy sequence. Let $\varepsilon \in \mathbb{Q}^+$. Since (a_n) is a Cauchy sequence, there is $N \in \mathbb{N}$ such that

$$|a_m - a_n| < e^2 \varepsilon, \quad \text{for all } m, n > N.$$

Hence, for all $m, n > \max(K, N)$, we have

$$|b_m - b_n| = \left| \frac{1}{a_m} - \frac{1}{a_n} \right|$$

$$= \frac{|a_n - a_m|}{|a_m a_n|}$$

$$= \frac{|a_n - a_m|}{|a_m||a_n|}$$

$$\leqslant \frac{|a_n - a_m|}{e^2}, \quad \text{since } |a_n| \geqslant e, \ |a_m| \geqslant e,$$

$$< \frac{e^2 \varepsilon}{e^2} = \varepsilon.$$

Last, we must show that $[a_n] \times [b_n] = [1]$.

$$a_n b_n = \begin{cases} 0 & \text{for } n \leqslant K \\ 1 & \text{for } n > K. \end{cases}$$

It is an easy exercise then to show that $(a_n b_n) \approx (1)$ (here (1) denotes the sequence with every term equal to 1) and, consequently, $[a_n b_n] = [1]$, as required.

Lemma

Let (a_n) be a Cauchy sequence in $\mathbb{Q}$ which does not have limit 0. Then there is a positive rational number e and there is a positive integer K such that $|a_n| \geqslant e$ for all $n > K$. (The sequence is eventually bounded away from zero.)

Proof

The negation of the statement '(a_n) has limit zero' is: there is $\varepsilon \in \mathbb{Q}^+$ such that, for every $N \in \mathbb{N}$, $|a_n - 0| \geqslant \varepsilon$ for *some* $n > N$. Since (a_n) is a Cauchy sequence there is $N_1 \in \mathbb{N}$ such that

$$|a_m - a_n| < \tfrac{1}{2}\varepsilon, \quad \text{for all } m, n > N_1.$$

Now, by the above, there is $K > N_1$ such that

$$|a_K| \geqslant \varepsilon.$$

Also, we have, for all $n > K$,

$$|a_n - a_K| < \tfrac{1}{2}\varepsilon.$$

It is now an exercise in manipulation of modulus signs to obtain the conclusion that

$$|a_n| \geqslant \tfrac{1}{2}\varepsilon, \quad \text{for all } n > K.$$

We therefore take e to be the number $\tfrac{1}{2}\varepsilon$, and the lemma is demonstrated.

▶ Next we turn to the ordering of the real numbers. We all have an intuitive feeling for relative sizes of real numbers, and for the geometric model of the real numbers in the real line. Our next purpose is to connect our newly-defined real numbers with these intuitions. As before, the first thing is to describe the set of *positive* numbers.

Definitions

(i) A Cauchy sequence (a_n) in $\mathbb{Q}$ is said to be *ultimately positive* if there exist a positive rational number e and a positive integer K such that $a_n \geq e$ for all $n > K$.

(ii) A real number $[a_n]$ is *positive* if the Cauchy sequence (a_n) is ultimately positive.

▶ For this definition to make sense we need to demonstrate a theorem.

Theorem 1.23

Let (a_n) and (b_n) be Cauchy sequences in $\mathbb{Q}$. If (a_n) is ultimately positive and $(a_n) \approx (b_n)$ then (b_n) is ultimately positive.

Proof

This is left as an exercise for the reader.

▶ The theorem ensures that if one Cauchy sequence is ultimately positive then every sequence in the corresponding equivalence class is ultimately positive.

$\mathbb{R}^+$ denotes the set of positive real numbers.

Definition

If $x, y \in \mathbb{R}$, we say that $x < y$ if $y - x \in \mathbb{R}^+$. Following the customary practice, $x \leq y$ means $x < y$ or $x = y$.

Remarks 1.24 (see Remarks 1.15)

(a) For $x, y, z \in \mathbb{R}$, we have $x < y$ if and only if $x + z < y + z$.

(b) For $x, y \in \mathbb{R}$ and $z \in \mathbb{R}^+$, we have $x < y$ if and only if $xz < yz$.

(c) For $x \in \mathbb{R}$ and $y \in \mathbb{R}^+$, we have $x < x + y$.

(d) If $x \in \mathbb{R}^+$ and $y \in \mathbb{R}^+$, then $x + y \in \mathbb{R}^+$.

(e) If $x \in \mathbb{R}^+$ and $y \in \mathbb{R}^+$, then $xy \in \mathbb{R}^+$.

(f) If $x \in \mathbb{R}^+$, then $-x < x$.

(g) $x \in \mathbb{R}^+$ if and only if $[0] < x$.

(h) For any $x \in \mathbb{R}$, $[0] \leq x^2$.

We sketch proofs for (a), (e) and (g). The others are left as exercises.

For (a), observe that $(y + z) - (x + z) = y - x$. Clearly, $(y + z) - (x + z) \in \mathbb{R}^+$ if and only if $y - x \in \mathbb{R}^+$, and the result follows.

For (e), let $x = [a_n] \in \mathbb{R}^+$, $y = [b_n] \in \mathbb{R}^+$. Then there exist $e_x, e_y \in \mathbb{Q}^+$ and $K_x, K_y \in \mathbb{N}$ such that $a_n \geq e_x$ for all $n > K_x$, and $b_n \geq e_y$ for all $n > K_y$. It follows that $a_n b_n \geq e_x e_y$ for all $n > \max(K_x, K_y)$ and, consequently, $(a_n b_n)$

is ultimately positive, as required. Notice that we are using here the corresponding property of $\mathbb{Q}$: from $e_x \in \mathbb{Q}^+$ and $e_y \in \mathbb{Q}^+$, we deduce that $e_x e_y \in \mathbb{Q}^+$.

For (g), we have only to note that, for any $x \in \mathbb{R}$, $x - [0] = x$. Thus $[0] < x$ if and only if $x - [0] \in \mathbb{R}^+$, i.e. $x \in \mathbb{R}^+$.

▶ The above shows that $\mathbb{R}$ has algebraic and order properties which are, in many respects, identical with those of $\mathbb{Q}$. We shall see that this also applies to some of the other properties of $\mathbb{Q}$ which were developed in the previous section. Of course, we shall eventually show that $\mathbb{R}$ also has the crucial property that $\mathbb{Q}$ does not have, namely, the least upper bound property.

Now, however, let us consider the embedding of $\mathbb{Q}$ in $\mathbb{R}$. We have had occasion already to consider the constant sequences $0, 0, 0, \ldots$ and $1, 1, 1, \ldots$, which give rise to the real numbers $[0]$ (zero) and $[1]$ (one). In fact, for any rational number a, the constant sequence $a, a, a, \ldots$ is a Cauchy sequence, and we denote the equivalence class it determines by $[a]$. In this way, to each $a \in \mathbb{Q}$ there corresponds an element $[a] \in \mathbb{R}$. It is easy to verify that we cannot have $[a] = [b]$ unless we have $a = b$ (in $\mathbb{Q}$), so it is a one–one correspondence. To see that the set $\{[a] \in \mathbb{R} : a \in \mathbb{Q}\}$ is a 'copy' of $\mathbb{Q}$ in $\mathbb{R}$, we must check the operations

$$[a] + [b] = [a + b],$$

and

$$[a] \times [b] = [ab].$$

Both hold, by the definitions of $+$ and $\times$ on $\mathbb{R}$. Moreover, the ordering is preserved: $a < b$ in $\mathbb{Q}$ holds if and only if $[a] < [b]$ in $\mathbb{R}$. Verification of this is left as an exercise.

Within $\mathbb{R}$ there is therefore an isomorphic copy of $\mathbb{Q}$ and, just as we have done in the previous constructions, we can regard the set $\{[a] \in \mathbb{R} : a \in \mathbb{Q}\}$ as a representation of the system of rational numbers. In practice we can forget the distinction between elements a of $\mathbb{Q}$ and elements $[a]$ of $\mathbb{R}$, and hence regard $\mathbb{Q}$ as a subset of $\mathbb{R}$.

Theorem 1.25

(i) Given $x \in \mathbb{R}$, *one* of the following holds:

$$x \in \mathbb{R}^+, x = 0, -x \in \mathbb{R}^+.$$

(ii) If $x, y \in \mathbb{R}^+$, then $x + y \in \mathbb{R}^+$ and $x \times y \in \mathbb{R}^+$.

Proof

(i) Let $x = [a_n] \in \mathbb{R}$. If $x \neq 0$ then the sequence (a_n) does not have limit 0. By the lemma given after the proof of Theorem 1.22, there are $e \in \mathbb{Q}^+$ and $K \in \mathbb{N}$ such that $|a_n| \geq e$ for all $n > K$. But (a_n) is a Cauchy sequence, so there exists $N \in \mathbb{N}$ such that

$$|a_m - a_n| < \tfrac{1}{2}e, \quad \text{for all } m, n > N.$$

Now fix $p > \max(K, N)$, so that $|a_p| \geq e$ and for all $n \geq p$

$$-\tfrac{1}{2}e < a_n - a_p < \tfrac{1}{2}e.$$

There are two cases to consider: $a_p \geq e$ and $a_p \leq -e$.

If $a_p \geq e$ then for all $n \geq p$ we have

$$a_n > a_p - \tfrac{1}{2}e \text{ by above,}$$

$$\geq e - \tfrac{1}{2}e = \tfrac{1}{2}e,$$

and in this case (a_n) is ultimately positive, so $x \in \mathbb{R}^+$.

Second, if $a_p \leq -e$, then for all $n \geq p$ we have

$$a_n < \tfrac{1}{2}e + a_p$$

$$\leq \tfrac{1}{2}e - e = -\tfrac{1}{2}e,$$

so that

$$-a_n \geq \tfrac{1}{2}e,$$

and in this case the sequence $(-a_n)$ is ultimately positive, so $-x \in \mathbb{R}^+$.

(ii) These have been given already as Remarks 1.24 (d) and (e).

Theorem 1.26

The natural ordering of $\mathbb{R}$ is dense, i.e. given $x, y \in \mathbb{R}$, with $x < y$, there is $z \in \mathbb{R}$ such that $x < z$ and $z < y$.

Proof

Take $z = (x + y)/2$. Details are left to the reader. Note that 2 is the real number [2], and division by 2 is multiplication by the inverse of [2], namely $[\tfrac{1}{2}]$.

Theorem 1.27

The natural ordering of $\mathbb{R}$ is Archimedean, i.e. given $x, y \in \mathbb{R}^+$, there is a positive integer r such that $y < rx$.

Proof

Let $x = [a_n]$ and $y = [b_n]$ be elements of $\mathbb{R}^+$, so that the sequences (a_n) and (b_n) are ultimately positive. We require to find a positive integer r such that $(ra_n - b_n)$ is ultimately positive. There exist $e \in \mathbb{Q}^+$ and $K \in \mathbb{N}$ such that $a_n \geq e$ for all $n > K$. Also, since (b_n) is a Cauchy sequence, it is bounded, i.e. there exists $d > 0$ in $\mathbb{Q}$ such that $|b_n| \leq d$ for every $n \in \mathbb{N}$.

Now $\mathbb{Q}$ is Archimedean (Theorem 1.18), and $e \in \mathbb{Q}^+$ and $d + 1 \in \mathbb{Q}^+$, so there is a positive integer r such that

$$d + 1 < re.$$

Hence, for all $n > K$, we have

$$d + 1 < re \leq ra_n.$$

Also, $b_n \leq |b_n| \leq d$ for every n, so we have

$$b_n + 1 < ra_n, \quad \text{for every } n > K,$$

and so

$$ra_n - b_n > 1, \quad \text{for every } n > K.$$

Consequently, the sequence $(ra_n - b_n)$ is ultimately positive, and the proof is complete. Notice that we have implicitly identified the positive integer r with the rational number r and with the real number r, according to our conventions.

Theorem 1.28

Given any non-empty subset A of $\mathbb{R}$ which is bounded above, there is a least upper bound in $\mathbb{R}$ for A. (This is usually expressed by saying that $\mathbb{R}$ has the least upper bound property.)

Proof

Let $A \subseteq \mathbb{R}$ and let $x \leq x_0$ for each $x \in A$, where $x_0 \in \mathbb{R}$. We construct equivalent Cauchy sequences (x_n) and (y_n) in $\mathbb{Q}$, which decrease and increase respectively, in order to 'trap' the least upper bound of A between them. First we find $a, b \in \mathbb{Q}$ such that the least upper bound (if it exists) lies between them. x_0 is an upper bound and would be suitable to be b if it were a member of $\mathbb{Q}$, but it may not be, so choose b to be some rational number greater than x_0 (there is such a number, by Theorem 1.27; see Exercise 10 at the end of this section). Similarly, choose a to be some rational number smaller than some element of A.

Now let us fix $n \in \mathbb{N}$ $(n \geqslant 1)$ for the moment. We have $b - a \in \mathbb{Q}^+$, so by Theorem 1.18 there is a positive integer r such that

$$b - a < r \times \frac{1}{n},$$

i.e.

$$a + \frac{r}{n} > b.$$

For such an r, the number $a + (r/n)$ is an upper bound for A. Hence, the set $\{r \in \mathbb{N} : a + (r/n) \text{ is an upper bound for } A\}$ is not empty, and by Theorem 1.5, it has a least member. Denote this least member by r_n.

Let

$$x_n = a + \frac{r_n}{n}$$

and

$$y_n = a + \frac{r_n - 1}{n}, \quad \text{for } n \in \mathbb{N}, n \geqslant 1.$$

Note that $x_n - y_n = 1/n$, so $y_n < x_n$, for each n. We can go further. Each x_n is an upper bound for A, while each y_n is *not*. Consequently,

$$y_m < x_n, \quad \text{for every } m, n.$$

Next we show that (x_n) and (y_n) are Cauchy sequences, and that $(x_n) \approx (y_n)$.

$$x_m - x_n < x_m - y_m \quad \text{since } y_m < x_n,$$

$$= \frac{1}{m}.$$

Also,

$$x_n - x_m < x_n - y_n \quad \text{since } y_n < x_m,$$

$$= \frac{1}{n}.$$

Therefore, for $m, n > k$, we have

$$|x_m - x_n| < \frac{1}{k}$$

and, consequently, (x_n) is a Cauchy sequence. The proof that (y_n) is a Cauchy sequence is similar. Moreover,

$$|x_n - y_n| = \frac{1}{n},$$

so for $n > k$ we have

$$|x_n - y_n| < \frac{1}{k},$$

and so

$$(x_n) \approx (y_n).$$

It remains to show that $[x_n]$ (which is equal to $[y_n]$) is the least upper bound for A in $\mathbb{R}$. First, suppose that it is not an upper bound, i.e. suppose that there is $[z_n] \in A$ with $[x_n] < [z_n]$. Then $(z_n - x_n)$ is ultimately positive, so there exist $e \in \mathbb{Q}^+$ and $K_1 \in \mathbb{N}$ such that $z_n - x_n \geqslant e$ for all $n > K_1$. Also, there exists $K_2 \in \mathbb{N}$ such that $|x_m - x_n| < \frac{1}{2}e$ for all $m, n > K_2$. Let $K = \max(K_1, K_2)$. Then, for any $n > K$, we have

$$|x_n - x_K| < \tfrac{1}{2}e, \quad \text{so that } x_n > x_K - \tfrac{1}{2}e,$$

and

$$z_n - x_n \geqslant e, \quad \text{so that } z_n \geqslant x_n + e.$$

Consequently,

$$z_n \geqslant x_K + \tfrac{1}{2}e,$$

and so

$$z_n - x_K \geqslant \tfrac{1}{2}e, \quad \text{for all } n > K.$$

Thus, comparing the sequence (z_n) with the constant sequence (x_K), we can see that $[x_K] < [z_n]$. But x_K is an upper bound for A (all terms in the sequence (x_n) are), and so $[z_n]$ must also be an upper bound for A. This contradicts our assumption about $[z_n]$, so we have shown that $[a_n]$ is an upper bound for A.

Finally, suppose that there is $[u_n] < [y_n]$ such that $[u_n]$ is an upper bound for A. We derive a contradiction again. $(y_n - u_n)$ is ultimately positive, so there exist $e \in \mathbb{Q}^+$ and $L_1 \in \mathbb{N}$ such that $y_n - u_n \geqslant e$ for all $n > L_1$. Also, there exists $L_2 \in \mathbb{N}$ such that $|y_m - y_n| < \frac{1}{2}e$, for all $m, n > L_2$. Let $L = \max(L_1, L_2)$. Then for any $n > L$, we have

$$|y_n - y_L| < \tfrac{1}{2}e, \quad \text{so that } y_L > y_n - \tfrac{1}{2}e,$$

and

$$y_n - u_n \geqslant e, \quad \text{so that } y_n \geqslant u_n + e.$$

Consequently,

$$y_L > u_n + \tfrac{1}{2}e,$$

and so

$$y_L - u_n > \tfrac{1}{2}e, \quad \text{for all } n > L.$$

Thus, as before, comparing the sequence (u_n) with the constant sequence (y_L), we see that $[y_L] > [u_n]$. But $[u_n]$ is an upper bound for A, so $[y_L]$ is also. This contradicts the construction of the sequence (y_n), since none of the terms y_n are upper bounds for A. This completes the proof.

▶ The set of real numbers is a field. Moreover, it is an ordered field (the relevant properties are those given in Theorem 1.25). Thus it is an ordered field with the least upper bound property. It can be shown, by a proof which is lengthy but not conceptually difficult, that any two ordered fields with the least upper bound property are isomorphic. We therefore have an algebraic way of characterising $\mathbb{R}$. Out of our construction of $\mathbb{R}$ from $\mathbb{N}$, through $\mathbb{Z}$ and $\mathbb{Q}$, has come a collection of basic properties of real numbers which serve to characterise the set $\mathbb{R}$ completely. For mathematicians who work in analysis, or with real numbers in some other area, the notion of $\mathbb{R}$ as an ordered field with the least upper bound property serves as an effective common starting point.

We referred earlier to inadequacies of the set of rational numbers. The absence of least upper bounds has been rectified in $\mathbb{R}$, but the other is still an inadequacy in $\mathbb{R}$. Not all polynomial equations with integer coefficients have solutions in $\mathbb{R}$. Certainly $x^2 - 2 = 0$ can be solved in $\mathbb{R}$ (though not, of course, in $\mathbb{Q}$), but the equation $x^2 + 2 = 0$ cannot be solved in $\mathbb{R}$. This leads to our last system of numbers, the complex numbers. Here we have a rather more straightforward construction than the others.

Definition

The set $\mathbb{C}$ of *complex numbers* is the set $\mathbb{R} \times \mathbb{R}$ of ordered pairs of real numbers. The operations of *addition* and *multiplication* are given by:

$$(a, b) + (c, d) = (a + c, b + d),$$

$$(a, b) \times (c, d) = (ac - bd, bc + ad).$$

▶ It is not technically difficult to show that these operations satisfy the usual laws of commutativity, associativity and distributivity. The element $(0, 0)$ behaves like zero, and the element $(1, 0)$ behaves like 1. The additive inverse of (a, b) is $(-a, -b)$, and the multiplicative inverse (provided a and b are not both zero) of (a, b) is $(a/(a^2+b^2), -b/(a^2+b^2))$. All these are easily verified.

So far, this notion of complex numbers may be unfamiliar. Where does $\sqrt{-1}$ fit in? Notice that

$$(0, 1) \times (0, 1) = (-1, 0) = -(1, 0).$$

Now $(1, 0)$ behaves like 1, so we have in effect here a square root of -1, namely $(0, 1)$. Let us denote $(0, 1)$ by i. Now, for any complex number (a, b) we can write

$$(a, b) = (a, 0) + (0, b),$$

and we can think of this as $a \times (1, 0) + b \times (0, 1)$, or as $a + bi$, with $a, b \in \mathbb{R}$. This is the customary notation for complex numbers. If we adopt this notation then the embedding of $\mathbb{R}$ in $\mathbb{C}$ becomes trivial. Complex numbers of the form $a + 0i$, (i.e. of the form $(a, 0)$) behave as real numbers do:

$$(a, 0) + (c, 0) = (a+c, 0),$$

and

$$(a, 0) \times (c, 0) = (ac, 0).$$

It is clear that the equations $x^2+2 = 0$ and $x^2+1 = 0$ can be solved in $\mathbb{C}$. Solutions are $x = \pm\sqrt{2}i$ and $x = \pm i$ respectively. It is rather a different proposition to demonstrate that *every* polynomial equation with integer coefficients has a solution in $\mathbb{C}$. This is often referred to as the fundamental theorem of algebra, and its proof requires methods which are not the concern of this book, so we shall omit it.

$\mathbb{C}$ is a field. The fundamental theorem of algebra holds in $\mathbb{C}$, but at the cost of losing the convenient order properties that $\mathbb{Q}$ and $\mathbb{R}$ have. $\mathbb{C}$ is *not* an ordered field, and there is no simple natural ordering of the elements of $\mathbb{C}$.

Exercises

1. Prove that if (x_n), (y_n), (x'_n), (y'_n) are Cauchy sequences in $\mathbb{Q}$ with $(x_n) \approx (x'_n)$ and $(y_n) \approx (y'_n)$, then $(x_n + y_n) \approx (x'_n + y'_n)$.
2. Show that the operations of addition and multiplication on $\mathbb{R}$ are commutative and associative.

3. Prove Theorem 1.23, i.e. that if (a_n) and (b_n) are Cauchy sequences in $\mathbb{Q}$, with (a_n) ultimately positive, such that $(a_n) \approx (b_n)$, then (b_n) is also ultimately positive.

4. Prove Remarks 1.24 (b), (c), (d), (f) and (h).

5. Let $a, b \in \mathbb{Q}$ be such that $(a) \approx (b)$. Prove that $a = b$.

6. Prove that, for every $a, b \in \mathbb{Q}$, $[a] < [b]$ in $\mathbb{R}$ if and only if $a < b$ in $\mathbb{Q}$.

7. Prove Theorem 1.26, i.e. that, for every $x, y \in \mathbb{R}$ with $x < y$, there exists $z \in \mathbb{R}$ with $x < z$ and $z < y$.

8. Let $x \in \mathbb{R}^+$. Show that there exists $q \in \mathbb{Q}^+$ with $q < x$.

9. (i) Let $x, y \in \mathbb{R}$ with $x < y$. Prove that there exists $z \in \mathbb{Q}$ such that $x < z$ and $z < y$.
   (ii) Let $x, y \in \mathbb{Q}$ with $x < y$. Prove that there exists $z \in \mathbb{R} \backslash \mathbb{Q}$ such that $x < z$ and $z < y$.

10. Let $x \in \mathbb{R}$. Show that there exist $a, b \in \mathbb{Q}$ with $a < x$ and $b > x$.

11. Prove that every Cauchy sequence in $\mathbb{R}$ has a limit in $\mathbb{R}$. (A sequence (x_n) of real numbers is a Cauchy sequence if, given any $\varepsilon > 0$ in $\mathbb{R}$, there is a positive integer N such that $|x_m - x_n| < \varepsilon$ for every $m, n > N$.)

12. Prove that if a sequence of elements of $\mathbb{Q}$ has a limit in $\mathbb{Q}$ then it is a Cauchy sequence. Does this result hold if we replace $\mathbb{Q}$ by $\mathbb{R}$ in either or both places?

13. (i) Prove that the set $\{a \in \mathbb{Q} : a^2 < 2\}$ has no least upper bound in $\mathbb{Q}$. Let x be the least upper bound in $\mathbb{R}$ for this set. Prove that $x^2 = 2$.
    (ii) Prove that every positive real number has a unique positive square root in $\mathbb{R}$.

14. Verify that addition and multiplication of complex numbers are commutative and associative. Also verify the distributive law.

1.4 Decimal notation

Let us now close this chapter with some remarks on the way that real numbers are normally represented. It is customary to use decimal notation when writing a number, for example

$$1.5, 3.333\ldots, 3.141\,59, 3.141\,592\,653\,5\ldots.$$

This notation has limitations, clearly, since the second and fourth examples do not specify real numbers at all, because they are not complete expressions. Normally we think of real numbers as decimal expressions which may not terminate. In practice, of course, it is impossible to write out fully a non-terminating decimal expression, and calculation with such expressions is rather difficult, except in special cases, for example where there is a recurring digit.

Let $n \in \mathbb{N}$ and let $a_1, a_2, \ldots$ be integers between 0 and 9 inclusive. What is meant by the expression

$$n \cdot a_1 a_2 a_3 \ldots ?$$

The best mathematical explanation of it is that it represents the sum of the series

$$n + \frac{a_1}{10} + \frac{a_2}{100} + \frac{a_3}{1000} + \cdots .$$

Alternatively, we can say that it represents the 'limit' of the sequence

$$n, \; n + \frac{a_1}{10}, \; n + \frac{a_1}{10} + \frac{a_2}{100}, \; n + \frac{a_1}{10} + \frac{a_2}{100} + \frac{a_3}{1000}, \ldots .$$

This sequence is a Cauchy sequence in $\mathbb{Q}$ (this was justified earlier), and the real number it determines is the number represented by the decimal expression.

Thus, to each decimal expression such as the above, there corresponds an element of $\mathbb{R}$. We should note here that although the expression $n \cdot a_1 a_2 a_3 \ldots$ is apparently non-terminating, it may happen that for all but finitely many suffixes i, a_i is zero. In this way the above remark covers both terminating and non-terminating decimals.

It requires a little more effort to demonstrate the converse, i.e. that each element of $\mathbb{R}$ may be represented by a decimal expression.

Theorem 1.29

Every positive real number may be represented uniquely by an expression $n \cdot a_1 a_2 a_3 \ldots$, where $n \in \mathbb{N}$ and each a_i is an integer with $0 \leq a_i \leq 9$, and where there is no $N \in \mathbb{N}$ such that $a_i = 9$ for all $i > N$ (i.e. the sequence $a_1, a_2, a_3, \ldots$ is not to end with an infinite sequence of 9's).

▶ Before we prove this theorem let us examine the reason for the stipulation about sequences of 9's. Consider the example $0.999 \ldots$. This represents the sum of the infinite series

$$\frac{9}{10} + \frac{9}{100} + \frac{9}{1000} + \cdots ,$$

i.e.

$$\frac{9}{10} \times \left(1 + \frac{1}{10} + \frac{1}{100} + \cdots \right),$$

i.e.

$$\frac{9}{10} \times \frac{1}{(1 - \frac{1}{10})}, \quad \begin{array}{l} \text{using the formula for the sum} \\ \text{of a geometric series,} \end{array}$$

i.e.

$$1.$$

Consequently, $1.000\ldots$ and $0.999\ldots$ are representations for the same number. By a similar argument we can show that $1.5000\ldots$ and $1.4999\ldots$ represent the same number, and likewise for any decimal expression ending with a recurring 9.

Proof (of Theorem 1.29)

Let $x \in \mathbb{R}^+$, so by the Archimedean property there is $r \in \mathbb{Z}^+$ such that $r \times 1 > x$. Take r to be the least such, and let $n = r - 1$, so that n is the greatest integer less than or equal to x. Write $x = n + a$, so that $a \in \mathbb{R}$ and $0 \le a < 1$.

Now consider the number $10a$. We know that $0 \le 10a < 10$. Let a_1 be the largest integer less than or equal to $10a$ (so that $0 \le a_1 < 10$). Then (say)

$$10a = a_1 + r_1, \quad \text{where } r_1 \in \mathbb{R} \text{ and } 0 \le r_1 < 1.$$

Consequently,

$$a = \frac{a_1}{10} + \frac{r_1}{10},$$

so that

$$x = n + \frac{a_1}{10} + \frac{r_1}{10}.$$

Repeat this process with the number $10r_1$, to obtain

$$10r_1 = a_2 + r_2,$$

so that

$$\frac{r_1}{10} = \frac{a_2}{100} + \frac{r_2}{100},$$

and

$$x = n + \frac{a_1}{10} + \frac{a_2}{100} + \frac{r_2}{100}, \quad \text{with } r_2 \in \mathbb{R} \text{ and } 0 \le r_2 < 1.$$

This process goes on, possibly indefinitely, and generates the digits $a_1, a_2, a_3, \ldots$ of the required decimal expression. Notice that the Archimedean property is used at each step.

The process cannot lead to a repeating 9. We illustrate by an example why this is so, leaving the proof to be filled in by the reader. Consider the number which could be represented by $3.7999\ldots$. As we have seen, this may also be represented as 3.800 . . . . Now in the above construction we would obtain $n = 3$, the greatest integer less than or equal to 3.8, and $a = 0.8$. Therefore $10a$ is 8, and a_1 is then 8, the greatest integer less than *or equal* to $10a$. The representation 3.7999 . . . thus cannot be the result of this construction.

Lastly, we demonstrate uniqueness. Suppose that $x = n \cdot a_1 a_2 a_3 \ldots = n' \cdot a_1' a_2' a_3' \ldots$. Now

$$\sum_1^\infty \frac{a_i}{10^i} < \sum_1^\infty \frac{9}{10^i} = 1,$$

since we do not have $a_i = 9$ for all i. Similarly,

$$\sum_1^\infty \frac{a_i'}{10^i} < 1.$$

Hence, the largest integer less than or equal to x is n, and is also n' and, consequently, $n = n'$. Now suppose that N is the smallest number such that $a_N \neq a_N'$, and suppose (without loss of generality) that $a_N' < a_N$. Then

$$0 = \sum_1^\infty \frac{a_i}{10^i} - \sum_1^\infty \frac{a_i'}{10^i}$$

$$= \sum_N^\infty \frac{a_i}{10^i} - \sum_N^\infty \frac{a_i'}{10^i}$$

$$= \frac{1}{10^N}(a_N - a_N') + \sum_{N+1}^\infty \frac{a_i}{10^i} - \sum_{N+1}^\infty \frac{a_i'}{10^i}$$

$$> \frac{1}{10^N} + 0 - \frac{1}{10^N},$$

since $a_N - a_N' \geq 1$ (both a_N and a_N' are integers), and since

$$\sum_{N+1}^\infty \frac{a_i'}{10^i} < \sum_{N+1}^\infty \frac{9}{10^i} < \frac{1}{10^N}.$$

Here we use the fact that sequence $a_1', a_2', a_3', \ldots$ does not end with a repeating 9. We thus have derived the contradiction $0 > 0$. Hence, we must have $a_i = a_i'$ for every i, and the two expressions for x are the same. This completes the proof of Theorem 1.29.

Remark 1.30

Negative real numbers have decimal representations derived from the corresponding positive real numbers in the normal manner.

For example the normal representation of $-\pi$ is $-3.141\,59\ldots$. Note that if the above construction were applied to $-\pi$ we would obtain $n = -4$ (the greatest integer less than or equal to $-\pi$), and $a_1 = 8$, $a_2 = 5$, $a_3 = 8$, etc. This would provide a perfectly reasonable way of representing negative real numbers, but it is not the normal one.

Let us consider the construction of Theorem 1.29 in relation to rational numbers.

Examples 1.31

(a) Let $x = \dfrac{15}{8}$. Then $n = 1$ and $a = \dfrac{7}{8}$. We obtain:

$$10a = \frac{70}{8} = 8 + \frac{6}{8}, \quad \text{so } r_1 = \frac{6}{8}.$$

$$10r_1 = \frac{60}{8} = 7 + \frac{4}{8}, \quad \text{so } r_2 = \frac{4}{8}.$$

$$10r_2 = \frac{40}{8} = 5 + 0, \quad \text{so } r_3 = 0.$$

Consequently, $r_i = 0$ for all $i \geq 3$, and we have

$$x = 1.875\,000\ldots.$$

(b) Let $x = \dfrac{2}{27}$. Then $n = 0$ and $a = \dfrac{2}{27}$. We obtain:

$$10a = \frac{20}{27} = 0 + \frac{20}{27}, \quad \text{so } r_1 = \frac{20}{27}.$$

$$10r_1 = \frac{200}{27} = 7 + \frac{11}{27}, \quad \text{so } r_2 = \frac{11}{27}.$$

$$10r_2 = \frac{110}{27} = 4 + \frac{2}{27}, \quad \text{so } r_3 = \frac{2}{27}.$$

$$10r_3 = \frac{20}{27} = 0 + \frac{20}{27}, \quad \text{so } r_4 = \frac{20}{27}.$$

From here on, the process repeats, and we obtain the infinite recurring expression

$$x = 0.074\,074\,074\ldots.$$

▶ As readers will no doubt be aware, terminating and recurring expressions have a special significance.

Theorem 1.32

A real number has a decimal expression which terminates or recurs if and only if it is a rational number.

Proof

First note that a terminating decimal expression is just one that ends with a recurring 0, so we deal in general just with the recurring case. Suppose, in the proof of Theorem 1.29, that the number x is rational, so that a is also rational, and let $a = \dfrac{p}{q}$ with $p, q \in \mathbb{N}$ and $q \neq 0$. Then

$$a = \frac{p}{q} = \frac{a_1}{10} + \frac{r_1}{10}.$$

It follows that

$$0 \leqslant r_1 = \frac{10p - a_1 q}{q} < 1,$$

so,

$$r_1 = \frac{p_1}{q}, \quad \text{say, with } 0 \leqslant p_1 < q.$$

Similarly,

$$r_2 = \frac{p_2}{q}, \quad \text{say, with } 0 \leqslant p_2 < q, \text{ etc.}$$

There must come a point when p_i is the same as p_j for some $j < i$, i.e. $r_i = r_j$. From this point on, the process will repeat and the sequence $a_j a_{j+1} \cdots a_{i-1}$ will recur in the decimal expression for x. This is clearly demonstrated in Example 1.31(b) above.

Now, for the converse, suppose that x has a recurring decimal expression, say

$$x = n \cdot a_1 a_2 \cdots a_r \dot{a}_{r+1} \cdots \dot{a}_{r+s},$$

where the $a_{r+1} \cdots a_{r+s}$ sequence recurs. Then

$$x = n + \frac{a_1}{10} + \cdots + \frac{a_r}{10^r} + \left(\frac{a_{r+1}}{10^{r+1}} + \cdots + \frac{a_{r+s}}{10^{r+s}} \right)$$

$$+\left(\frac{a_{r+1}}{10^{r+s+1}}+\cdots+\frac{a_{r+s}}{10^{r+2s}}\right)+\left(\qquad\right)+\cdots$$

$$=n+\frac{a_1}{10}+\cdots+\frac{a_r}{10^r}+\left(\frac{a_{r+1}}{10^{r+1}}+\cdots+\frac{a_{r+s}}{10^{r+s}}\right)$$

$$+\frac{1}{10^s}\left(\frac{a_{r+1}}{10^{r+1}}+\cdots+\frac{a_{r+s}}{10^{r+s}}\right)+\frac{1}{10^{2s}}\left(\qquad\right)+\cdots$$

$$=n+\frac{a_1}{10}+\cdots+\frac{a_r}{10^r}$$

$$+\left(\frac{a_{r+1}}{10^{r+1}}+\cdots+\frac{a_{r+s}}{10^{r+s}}\right)\left(1+\frac{1}{10^s}+\frac{1}{10^{2s}}+\cdots\right)$$

$$=n+\frac{a_1}{10}+\cdots+\frac{a_r}{10^r}$$

$$+\left(\frac{a_{r+1}}{10^{r+1}}+\cdots+\frac{a_{r+s}}{10^{r+s}}\right)\times\frac{1}{1-\dfrac{1}{10^s}}\ ,$$

and this is a rational number.

Remarks 1.33

(a) It is possible to *define* real numbers to be decimal expressions as above, rather than equivalence classes of Cauchy sequences as we have done. Much of the development would be quite similar – some properties would be easier to derive, and others more complicated. One of the principal difficulties would be how to define the product of two such expressions.

(b) The number 10 occurs because decimal representation is a historical fact. All of the above development can be carried through in other bases with only trivial modifications. The details of this are, again, part of number theory, so we shall not pursue them here.

Exercises

1. $2.569\,99\ldots$ and 2.57 are decimal representations of the same real number. Work through the last part of the proof of Theorem 1.29 to see precisely why a contradiction cannot be derived by that procedure from the equation $2.5699\ldots = 2.57$.

2. By the method of Theorem 1.29, find decimal expansions for the rational numbers 1/7, 2/9, 3/23. Can you suggest any general rule governing the length of the recurring sequence in such expansions?

Further reading

Beth [2] A large scale, detailed and sophisticated exposition of the foundations of mathematics, including much about philosophical matters.

Bostock [4], [5] The basis and development of numbers from a more philosophical point of view than ours (including quite a lot of logic).

Dedekind [7] The views of the originator of these ideas, though somewhat obscure by today's standards.

Kline [17] A huge work, encompassing all of the history and development of mathematics.

Mendelson [19] All the details of the constructions of the number systems, including the basic algebraic ideas and the algebra of sets.

2

THE SIZE OF A SET

Summary

This chapter is concerned with relative sizes of sets, through the idea of functions between them. The distinctions between finite and infinite sets, and between countable sets and uncountable sets, are made. Properties of countable sets are derived, and the sets $\mathbb{Z}$ and $\mathbb{Q}$ are shown to be countable. $\mathbb{R}$ is shown to be uncountable, and properties of sets equinumerous with $\mathbb{R}$ are derived. Two sets are said to have the same cardinal number if there is a bijection between them. Properties of the cardinal numbers $\aleph_0$ and $\aleph$ are derived.

The reader is presumed to be familiar with the algebra of sets and with the notions of injection, surjection and bijection. Apart from one reference to Theorem 1.29, this chapter is independent of Chapter 1, although knowledge of the basic properties of integers, rational numbers and real numbers is required.

2.1 Finite and countable sets

How can we measure the size of a set? Perhaps the crudest criterion is that of finiteness. A set is either finite or infinite, and the former is 'smaller' than the latter. For finite sets there is an obvious further measure of size, namely, the number of elements in the set, and using this criterion it is easy to judge when one finite set is 'larger' than another. For infinite sets the question is not so easy however. This book is largely about the mathematical ideas necessary for sensible discussion of the nature and behaviour of infinite sets. In this chapter we concentrate on the less formal side of this and present some basic results about sizes of infinite sets. This we can do without an axiomatic approach, and it is best so done, but we shall find that our intuition can take us only so far – that there are certain difficulties relating to some apparently innocent procedures, and that some simple properties have far reaching consequences. The axiomatic approach comes in later to provide a framework for sorting out the interdependences between certain principles. The prime example of this concerns the *axiom of choice*, which we shall see has several apparently unconnected consequences, some of which are intuitively acceptable and others perhaps less so.

The elements of a finite set can be counted. This counting process is an association between the positive integers from 1 to n (say) and the elements of the given set, thus:

$$1, \quad 2, \quad 3, \quad \ldots, n.$$

$$a_1, a_2, a_3, \ldots, a_n.$$

This association may be regarded as a function from the set $\{1, 2, \ldots, n\}$ to the given set. It is not merely a function, though. It is a bijection, i.e. it is a one-one and onto function. It is one-one because we ensure that no element of the set is counted twice, and it is onto because we count the whole set. Thus a set A has n elements if and only if there is a bijection from $\{1, \ldots, n\}$ to A.

Definition
A non-empty set A is *finite* if there is a positive integer n and a bijection from $\{1, \ldots, n\}$ to A. Otherwise it is *infinite*. The empty set is by convention taken to be finite.

Definition
Two sets A and B are *equinumerous* if there is a bijection from A to B. We denote this by $A \sim B$.

This definition is based on the consideration of the sizes of finite sets, but it may be applied equally to infinite sets, as we shall see.

The relation of equinumerosity between sets has the following straightforward properties.

Theorem 2.1
For any sets A, B and C:
(i) $A \sim A$.
(ii) If $A \sim B$, then $B \sim A$.
(iii) If $A \sim B$ and $B \sim C$, then $A \sim C$.

Proof

(i) The identity function is a bijection from A to A, so $A \sim A$.

(ii) Let $f: A \to B$ be a bijection. Then $f^{-1}: B \to A$ exists and is a bijection.

(iii) Let $A \sim B$ and $B \sim C$. Then bijections $g: A \to B$ and $h: B \to C$ exist. Hence $A \sim C$, since $f \circ g$ is a bijection from A to C.

Example 2.2

Consider the sets $\mathbb{N} = \{0, 1, 2, \ldots\}$ and $2\mathbb{N} = \{0, 2, 4, \ldots\}$. The second is a proper subset of the first. Is it a 'smaller' set? In one sense clearly it is, for $\mathbb{N}$ contains elements which are not contained in $2\mathbb{N}$. However, $\mathbb{N}$ and $2\mathbb{N}$ are *equinumerous*. The function $f: \mathbb{N} \to 2\mathbb{N}$ given by $f(x) = 2x$ $(x \in \mathbb{N})$ is a bijection.

► The above example shows that an infinite set can be equinumerous with a proper subset of itself, a situation which is clearly impossible for finite sets. Indeed, the property of having a proper subset equinumerous with the whole set has been proposed as a definition of infiniteness (Dedekind infiniteness). However, we shall not pursue this here as this is one area where the axiom of choice unavoidably comes in. We shall see many examples of sets with equinumerous subsets, and shall return to this matter in Chapter 5.

Examples 2.3

(a) $g: \mathbb{R} \to \mathbb{R}^+$, given by $g(x) = e^x$ $(x \in \mathbb{R})$, is a bijection, so $\mathbb{R}$ is equinumerous with $\mathbb{R}^+$.

(b) $h: \mathbb{R} \times \mathbb{R} \to \mathbb{C}$, given by $h(x, y) = x + iy$, is a bijection, so $\mathbb{R} \times \mathbb{R}$ is equinumerous with $\mathbb{C}$.

► Now that we have clarified what we shall mean by 'having the same size' let us now turn to the notions 'smaller than' and 'larger than'. In our terms we cannot regard $2\mathbb{N}$ as strictly smaller than $\mathbb{N}$, because these sets are equinumerous. However, the finite set $\{1, 2, 3, 4, 5\}$ is certainly a smaller set than $\mathbb{N}$. More generally, if A is any finite set then A is equinumerous with $\{1, 2, \ldots, n\}$, for some n, and this set is smaller than $\mathbb{N}$ (because it is contained in $\mathbb{N}$ and not equinumerous with $\mathbb{N}$). We therefore say that A is strictly dominated by $\mathbb{N}$, introducing a new word for this restricted and precise concept.

$\leq$ injal
$\geqslant \neq$ Surye
$\sim$ Bijel

Definition

For sets A and B, A is *dominated* by B if there is an injection (one-one function) from A to B. We write $A \leqslant B$. A is *strictly* dominated by B if $A \leqslant B$ and A is not equinumerous with B.

Theorem 2.4

(i) If A is a finite set, then $A \leqslant \mathbb{N}$.

(ii) For any sets A and B, if $A \sim B$, then $A \leqslant B$.

(iii) For any set A, $A \leqslant A$.

(iv) For any sets A and B, if $A \subseteq B$, then $A \leqslant B$.

Proof

(i) Let A have n elements. Then $A \sim \{1, 2, \ldots, n\}$, so there is a bijection $A \to \{1, 2, \ldots, n\}$, which can be regarded as an injection $A \to \mathbb{N}$. Hence $A \leqslant \mathbb{N}$.

(ii) Let $A \sim B$ via a bijection f. Certainly, f is an injection, so trivially $A \leqslant B$.

(iii) This is an immediate consequence of (ii).

(iv) If $A \subseteq B$ then let $f: A \to B$ be given by $f(a) = a$ $(a \in A)$. Then f is an injection from A to B, so $A \leqslant B$. (This function f is known as the *inclusion* function from A to B.)

Theorem 2.5

If $A \leqslant B$ and $C \sim A$, then $C \leqslant B$.

Proof

Exercise.

▶ The elements of a finite set can be counted, as we have seen. Now we shall extend the notion of counting, as follows. An infinite set X is to be 'countable' if there is an association between the set of all non-negative integers and X:

$$0, \ 1, \ 2, \ 3, \ \ldots$$

$$x_0, x_1, x_2, x_3, \ldots.$$

More precisely, we require the existence of a rule for listing the elements of X without repetition in such a way that each element of X occurs at some finite point in the list. For example, the set $2\mathbb{N}$ is countable in this sense, because the list $0, 2, 4, 6, \ldots$ fulfils the requirement.

Some readers may find the idea of an infinite list rather vague, so let us be still more precise. The existence of such an infinite list is equivalent to the existence of a bijection between $\mathbb{N}$ and X, for if $f:\mathbb{N}\to X$ is a bijection then $f(0)$, $f(1)$, $f(2)$, ... is a list without repetitions containing all members of X, and if x_0, x_1, x_2, x_3, ... is such a list then $g:\mathbb{N}\to X$ given by $g(n)=x_n$ is a bijection. We therefore make the following definition.

Definition

A set A is *countable* if either
(a) it is finite, or
(b) it is infinite and $\mathbb{N}\sim A$.

Examples 2.6

(a) $\mathbb{N}$ itself is countable.
(b) All subsets of $\mathbb{N}$ are countable. To see this, let A be a subset of $\mathbb{N}$. If A is finite then it is certainly countable. If A is infinite, we can make a list of its elements by listing $\mathbb{N}$ (in order), deleting as we proceed all members of $\mathbb{N}\setminus A$. We obtain a listing of the elements of A.
(c) Any subset of a countable set is countable. A proof of this is similar to that of (b), and is left as an exercise.
(d) The set of all complex roots of unity is countable. For each $n \geqslant 1$ there are n complex roots of unity, and we can make a list of all the roots by writing down 1, then -1, then the two cube roots other than 1, then the two fourth roots other than 1 and -1, and so on, omitting all repetitions as we go.
(e) If A is a countable set and $A \sim B$, then B is countable.

▶ A notion such as countability would be pointless if it were a property of all sets, and it is certainly not clear yet whether all sets are indeed countable. After some further theorems we shall be able to see that sets exist which are not countable and to derive some results about them, but first let us examine countable sets and their properties.

The definition of countability is somewhat cumbersome to apply. The next two results yield more convenient criteria for deciding countability of sets.

Theorem 2.7

A set A is countable if and only if there is an injection $A \to \mathbb{N}$ (i.e. $A \leqslant \mathbb{N}$).

Proof

First suppose that A is countable. If A is finite, say $A = \{a_1, \ldots, a_n\}$, then the function which maps a_k to k $(1 \le k \le n)$ is an injection $A \to \mathbb{N}$. If A is infinite then by the definition of countability there is a bijection $\mathbb{N} \to A$, whose inverse is an injection $A \to \mathbb{N}$.

Now suppose that there is an injection $h : A \to \mathbb{N}$. Then $h(A)$ is a subset of $\mathbb{N}$, so $h(A)$ is countable (see Example 2.6(b)). But $h : A \to h(A)$ is a bijection since h is an injection. Either $h(A)$ is finite, in which case A must also be finite (and hence countable) or $h(A)$ is equinumerous with $\mathbb{N}$. In the latter case we have $A \sim h(A)$ and $h(A) \sim \mathbb{N}$, so $A \sim \mathbb{N}$ by Theorem 2.1(iii), and A is countable.

Corollary 2.8

A non-empty set A is countable if and only if there is a surjection $\mathbb{N} \to A$.

Proof

Let A be a countable non-empty set. By the theorem there is an injection $f : A \to \mathbb{N}$. Then f is a bijection between A and $f(A)$, so there is an inverse bijection f^{-1} from $f(A)$ to A. Choose an element $a_0 \in A$. Define $g : \mathbb{N} \to A$ by

$$g(n) = \begin{cases} f^{-1}(n) & \text{if } n \in f(A) \\ a_0 & \text{if } n \notin f(A). \end{cases}$$

Since f is a bijection between A and $f(A)$, g is a surjection onto A.

Now suppose that there is a surjection $g : \mathbb{N} \to A$. Define $f : A \to \mathbb{N}$ by

$$f(a) = \text{smallest } n \in \mathbb{N} \text{ such that } g(n) = a.$$

Then f is an injection as required.

▶ This theorem and corollary will be very useful, since, in each of them, the separate cases contained in the definition of countability are subsumed under a single necessary and sufficient condition. We shall use them repeatedly in the derivation of properties of countable sets and in our proofs that certain familiar sets are countable.

Theorem 2.9

The union of two countable sets is countable.

Proof

Let A and B be countable sets and let $f:\mathbb{N}\to A$ and $g:\mathbb{N}\to B$ be surjections. Define $h:\mathbb{N}\to A\cup B$ by

$$h(n)=\begin{cases}f(k) & \text{if } n=2k+1 \ (k\in\mathbb{N})\\ g(k) & \text{if } n=2k \ (k\in\mathbb{N}).\end{cases}$$

h is clearly a surjection, so $A\cup B$ is countable.

Corollary 2.10

The union of any finite collection of countable sets is countable.

Proof

The proof is by induction on the number n of sets in the collection. Let the sets be denoted by $A_1, A_2, \ldots, A_n$, where $n\geqslant 2$.
Base step: $n=2$. This is just Theorem 2.9.
Induction step: Let $n>2$. Suppose that the union of a collection of $n-1$ countable sets is countable, so that $(A_1\cup\cdots\cup A_{n-1})$ is countable. Then $A_1\cup A_2\cup\cdots\cup A_n=(A_1\cup\cdots\cup A_{n-1})\cup A_n$, a union of two countable sets, which is countable by Theorem 2.9.

Corollary 2.11

The set $\mathbb{Z}$ of integers is countable.

Proof

$\mathbb{Z}^+$ is countable (being a subset of $\mathbb{N}$). The set $\mathbb{Z}^-$ of negative integers is countable, since it is equinumerous with $\mathbb{Z}^+$ (the function which maps $n\to -n$ is a bijection $\mathbb{Z}^+\to\mathbb{Z}^-$). Also, the set $\{0\}$ is countable, since it is finite.

$$\mathbb{Z}=\mathbb{Z}^+\cup\{0\}\cup\mathbb{Z}^-,$$

so $\mathbb{Z}$ is a union of three countable sets and thus, by Corollary 2.10, is countable.

Remark

It is certainly possible to prove that $\mathbb{Z}$ is countable by direct construction of a bijection between $\mathbb{Z}$ and $\mathbb{N}$. The reader is recommended to verify that the function f defined as follows is a bijection from $\mathbb{Z}$ to $\mathbb{N}$.

$$f(x)=\begin{cases}2|x| & \text{if } x\geqslant 0\\ 2|x|+1 & \text{if } x<0.\end{cases}$$

▶ To demonstrate countability of a set S we may, by Theorem 2.7, find an injection from S into $\mathbb{N}$. One very convenient device for doing this is the use of products of primes or powers of primes. The fact that any positive integer can be *uniquely* expressed (apart from the order of the factors) as a product of primes is what ensures that our functions are injections.

(This fact is known as the fundamental theorem of arithmetic, and a proof of it may be found in any textbook on elementary number theory.)

The proof of the next theorem will illustrate this procedure.

Theorem 2.12

The Cartesian product of two countable sets is a countable set.

Proof

Let A and B be countable sets, and suppose that $f : A \to \mathbb{N}$ and $g : B \to \mathbb{N}$ are injections. Then $h : A \times B \to \mathbb{N}$ is an injection, where

$$h(a, b) = 2^{f(a)} \times 3^{g(b)} \quad (a \in A, b \in B).$$

To see this let (a, b) and (a', b') be elements of $A \times B$ with $h(a, b) = h(a', b')$. Then

$$2^{f(a)} \times 3^{g(b)} = 2^{f(a')} \times 3^{g(b')}.$$

By the uniqueness of prime power decomposition, then, we must have $f(a) = f(a')$ and $g(b) = g(b')$. Since f and g are injections, $a = a'$ and $b = b'$, and so $(a, b) = (a', b')$. Hence, h is an injection, as required, and $A \times B$ is countable.

▶ This method can be extended as follows.

Theorem 2.13

The Cartesian product of any finite number of countable sets is a countable set.

Proof

Let $A_0, \ldots, A_n$ be countable sets, and let $f_i : A_i \to \mathbb{N} \ (0 \leq i \leq n)$ be injections. Denote by $p_0, p_1, p_2, \ldots$ the sequence of prime numbers in order of magnitude and define $f : A_0 \times \cdots \times A_n \to \mathbb{N}$ by:

$$f(a_0, a_1, \ldots, a_n) = 2^{f_0(a_0)} \times 3^{f_1(a_1)} \times \cdots \times p_n^{f_n(a_n)}.$$

As in the previous proof we can show that f is an injection, so $A_0 \times \cdots \times A_n$ is countable.

▶ So finite unions and finite Cartesian products of countable sets yield countable sets. With a little sleight of hand we can deal also with countable infinite unions.

Theorem 2.14

The union of a countable collection of countable sets is countable.

Proof

We give two proofs. The first involves less formal considerations and may help to give an intuitive grasp of the ideas of this chapter. The second is a more rigorous version, based on the same process.

Consider the countable sets $A_0, A_1, A_2, \ldots$ (not necessarily pairwise disjoint). These sets may be listed, say as follows:

$$A_0 = \{a_{00}, a_{01}, a_{02}, \ldots \},$$

$$A_1 = \{a_{10}, a_{11}, a_{12}, \ldots \},$$

$$A_2 = \{a_{20}, a_{21}, a_{22}, \ldots \},$$

$$A_3 = \{a_{30}, a_{31}, a_{32}, \ldots \},$$

etc.

We obtain an infinite array which contains all the elements of $\bigcup_{i \in \mathbb{N}} A_i$ (possibly with repetitions). All of the entries in this array may be put in a single infinite list by starting thus:

$$a_{00}, a_{10}, a_{01}, a_{02}, a_{11}, a_{20}, a_{30}, a_{21}, a_{12}, a_{03}, a_{04}, \ldots$$

and following successively each diagonal across the array. In this list we can now delete all repetitions, and what remains is a list of all the elements of $\bigcup_{i \in \mathbb{N}} A_i$, so $\bigcup_{i \in \mathbb{N}} A_i$ is a countable set.

Now for the more formal proof. Let $A = \bigcup_{i \in \mathbb{N}} A_i$. By Corollary 2.8 there exists a surjection $f_i : \mathbb{N} \to A_i$ for each $i \in \mathbb{N}$. We construct a surjection f from $\mathbb{N}$ to A, and the result will follow by Corollary 2.8. Given $n \in \mathbb{N}$, if $n \neq 0$ we may write $n = 2^k \times 3^l \times m$, where $m \neq 0$ and m is not divisible by 2 or 3, and this expression is unique. So let

$$\begin{cases} f(0) = f_0(0) \\ f(n) = f_k(l) & \text{if } n = 2^k \times 3^l \times m \text{ as above.} \end{cases}$$

To see that f is a surjection, let $x \in A$. Then $x \in A_i$ for some $i \in \mathbb{N}$, so $x = f_i(r)$ for some $r \in \mathbb{N}$. But then $x = f(2^i \times 3^r)$.

▶ The reader who is familiar with the axiom of choice may care to consider where in the above proof the axiom of choice is (implicitly) used. This result requires the assumption of at least a weak form of the axiom of choice (and this is the 'sleight of hand' referred to above). We shall return to this in Chapter 5. See Theorem 5.21.

Similar ideas to those used in the above proofs are used to obtain what is perhaps a surprising result about the set of rational numbers.

Theorem 2.15
$\mathbb{Q}$ is a countable set.

Proof
Construct an injection $f : \mathbb{Q}^+ \to \mathbb{N}$ as follows. Given $x \in \mathbb{Q}^+$, there exist uniquely determined positive integers p and q such that $x = p/q$ and p and q have no common divisor greater than 1. Let $f(x) = 2^p \times 3^q$. Verification that f is an injection is again left as an exercise. Hence $\mathbb{Q}^+$ is countable. $\mathbb{Q}^-$ is therefore countable also (it is clearly equinumerous with $\mathbb{Q}^+$). Now $\mathbb{Q} = \mathbb{Q}^+ \cup \{0\} \cup \mathbb{Q}^-$, a union of three countable sets, so $\mathbb{Q}$ is countable, by Corollary 2.10.

Remark
Theorem 2.15 may be proved in a less formal way by constructing an array containing all positive rational numbers thus:

$$\frac{1}{1}, \frac{1}{2}, \frac{1}{3}, \frac{1}{4}, \ldots,$$

$$\frac{2}{1}, \frac{2}{2}, \frac{2}{3}, \frac{2}{4}, \ldots,$$

$$\frac{3}{1}, \frac{3}{2}, \frac{3}{3}, \frac{3}{4}, \ldots,$$

etc.,

and proceeding as in the previous situation to obtain a single list containing all elements of $\mathbb{Q}^+$ (without repetitions), thus demonstrating the countability of $\mathbb{Q}^+$. The remainder of the proof is then as above.

► Countable sets abound, and it is not clear yet how non-countable sets might be constructed. In the next section we shall be dealing with comparison of the sizes of infinite sets in a more general way than hitherto, so let us complete this section with a demonstration that non-countable sets exist.

Definition
Given a set A, the set of all subsets of A is called the *power set* of A, and denoted by $P(A)$.

► What we seek is a way of obtaining a set which is 'bigger' than $\mathbb{N}$, sufficiently so that it is not equinumerous with $\mathbb{N}$. Note that, for any set A, there is an injection $A \to P(A)$ which maps each element a to the singleton subset $\{a\}$, so $A \preccurlyeq P(A)$. However $P(A)$ contains many members other than the singletons, and it turns out that A and $P(A)$ cannot be equinumerous.

Theorem 2.16
For any set A, there is no bijection from A to $P(A)$.

Proof
Suppose the contrary, and let $f : A \to P(A)$ be a bijection. Then for each $x \in A$, $f(x)$ is a subset of A, of which x itself may or may not be an element. Let

$$T = \{x \in A : x \notin f(x)\}.$$

T is a subset of A, i.e. $T \in P(A)$, and since f is a bijection we must have $T = f(y)$ for some $y \in A$. Now either $y \in T$ or $y \notin T$. If $y \in T$ then $y \notin f(y)$ by the definition of T, i.e. $y \notin T$ (contradiction). Also, if $y \notin T$ then $y \notin f(y)$, since $T = f(y)$, so y satisfies the requirement for belonging to T, i.e. $y \in T$ (contradiction). Either way we obtain a contradiction, so the original assumption must have been false. Therefore there cannot exist any set A with a bijection $f : A \to P(A)$.

Corollary 2.17
$P(\mathbb{N})$ is not a countable set.

Proof
$P(\mathbb{N})$ is clearly infinite. By Theorem 2.16 there is no bijection between $\mathbb{N}$ and $P(\mathbb{N})$. Hence $P(\mathbb{N})$ is not countable.

Definition

A set is *uncountable* if it is infinite and it is not equinumerous with $\mathbb{N}$.

Exercises

1. Prove that the following functions are bijections and hence that the two sets in each case are equinumerous.
   (i) $f:\mathbb{R}\to\mathbb{R}^+\cup\{0\}$ given by $f(x)=x^2$.
   (ii) $f:\mathbb{Z}\to\mathbb{N}$ given by $f(x)=\begin{cases}2|x| & \text{if } x\geqslant 0 \\ 2|x|+1 & \text{if } x<0.\end{cases}$
   (iii) $f:\mathbb{N}\times\mathbb{N}\to 2\mathbb{N}\times 3\mathbb{N}$ given by $f(x,y)=(2x,3y)$.
   (iv) $f:\mathbb{R}\to\mathbb{R}^+$ given by $f(x)=e^x$.
   (v) $f:[0,1]\to[1,3]$ given by $f(x)=2x+1$.
   (vi) $f:\mathbb{R}^+\times[0,2\pi)\to\mathbb{C}\backslash\{0\}$ given by $f(x,y)=x\,e^{iy}$.
   (vii) $f:\left(-\dfrac{\pi}{2},\dfrac{\pi}{2}\right)\to\mathbb{R}$ given by $f(x)=\tan x$.

2. In each case below, by describing a procedure for generating a list of the elements, show that the given set is countable.
   (i) All words in the English language.
   (ii) All sentences in the English language.
   (iii) All matrices with entries from the set $\{0,1\}$.
   (iv) All 2×2 matrices with positive integer entries.
   (v) All 3-element subsets of $\mathbb{N}$.

3. By finding injections into $\mathbb{N}$, show that the following sets are countable.
   (i) Any infinite cyclic group.
   (ii) All 2×2 matrices with entries from $\mathbb{N}$.
   (iii) All 2×2 matrices with entries from $\mathbb{Z}$.
   (iv) All 4-element subsets of $\mathbb{N}$.
   (v) All 4-element sequences of elements of $\mathbb{N}$.

4. For each pair of sets A, B given below, find whether $A\leqslant B$ or $B\leqslant A$ (or both).
   (i) $A=\mathbb{N}$, $B=\mathbb{Z}$.
   (ii) $A=\mathbb{N}$, $B=$ any finite set.
   (iii) $A=\mathbb{N}$, $B=\boldsymbol{P}(\mathbb{N})$.
   (iv) $A=\mathbb{N}\times\mathbb{N}$, $B=\mathbb{Q}\times\mathbb{Q}$.
   (v) $A=[0,1]$, $B=[0,2]$ (intervals in $\mathbb{R}$).
   (vi) $A=\boldsymbol{P}(\mathbb{N})$, $B=\mathbb{Z}\times\mathbb{Z}$.

5. Show that $\leqslant$ is reflexive and transitive, i.e. that $A\leqslant A$ for any set A, and that $A\leqslant B$ and $B\leqslant C$ imply $A\leqslant C$.

6. Prove that if A is an infinite countable set then $A\times A\sim A$.

7. Prove that the set of all non-constant polynomials with integer coefficients is countable. Deduce that the set of all complex numbers which are roots of such polynomials is countable. What can be said about the set of all real numbers which are roots of such polynomials?

8. Show that the set of all *finite* subsets of $\mathbb{N}$ is countable, and deduce that the set of all infinite subsets of $\mathbb{N}$ is not countable.

9. Let A be an infinite set with an infinite countable subset B. Let X be a countable set. Prove that $A \cup X$ is equinumerous with A. (Hint: $B \cup X$ is equinumerous with B.)

10. (i) Let X be a finite set, with n elements. How many elements does $P(X)$ contain?
    (ii) Does there exist a set Y such that $P(Y)$ is infinite and countable?

11. Let A be a set which contains a proper subset B such that $A \sim B$. Show that A contains a subset which is equinumerous with $\mathbb{N}$ and that A is therefore infinite.

12. Where is the fallacy in the following argument? The set of rational numbers is countable and dense in $\mathbb{R}$ (so that given any real number, there are rational numbers arbitrarily close to it). Let $q_1, q_2, q_3, \ldots$ be an enumeration of the elements of $\mathbb{Q}$, and for each q_n, let I_n be the interval with midpoint q_n and length $1/2^n$. $I_1 \cup I_2 \cup I_3 \cup \cdots$ contains each rational number and a finite interval around it, so is all of $\mathbb{R}$, by the density of $\mathbb{Q}$ in $\mathbb{R}$. However, the sum of the lengths of the intervals I_n is $\sum_{n=1}^{\infty} 1/2^n$, i.e. 1, so the real line $(=I_1 \cup I_2 \cup \cdots)$ has length at most 1.

2.2 Uncountable sets

A demonstration that two given sets are equinumerous requires verification that a bijection exists. This can be done by direct or indirect methods. However, construction of a bijection can be difficult – the hard part often is ensuring that the function constructed is a surjection. An example of an indirect method is: if we can show that the two given sets are infinite and countable, then we know that they are equinumerous. This example requires (by Theorem 2.7) only the construction of two *injections*, and injections tend to be rather easier to find than bijections. This leads us to a very useful and significant result, which is far from obvious but whose proof is elementary in nature. It was first proved around the turn of the century, and it is still known by the names of the discoverers of its first proof.

Theorem 2.18 (the Schröder–Bernstein theorem)

If A and B are sets and there exist injections $f: A \to B$ and $g: B \to A$ then there exists a bijection between A and B. (Equivalently: if $A \leqslant B$ and $B \leqslant A$ then $A \sim B$.)

Proof

The proof is lengthy, and its methods are not used subsequently so the reader may omit it on first reading without loss of understanding.

We have injections $f: A \to B$ and $g: B \to A$. Consider any $b_1 \in B$. Let us attempt to construct a sequence $b_1, a_1, b_2, a_2, b_3, \ldots$ of alternating elements of A and B in the following way. First, there may or may not exist $a_1 \in A$ such that $f(a_1) = b_1$, but if such does exist, it is unique, since f is an injection. So we choose a_1 to be the inverse image of b_1 (under f), if it exists (see Fig. 2.1). Supposing that we have obtained a_1, we

Fig. 2.1

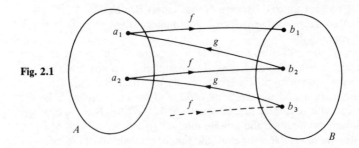

choose b_2 to be the unique element of B such that $g(b_2) = a_1$. Again, there may not be any such element, but if there is one, it is unique, since g is an injection. Similarly, we choose a_2 to be the inverse image of b_2 (under f), if it exists, and so on. If we continue this process as far as possible, one of three things must happen:

(1) We reach some $a_n \in A$ and stop because there is no $b^* \in B$ with $g(b^*) = a_n$. This situation is possible because g need not be a surjection.

(2) We reach some $b_n \in B$ and stop because there is no $a^* \in A$ with $f(a^*) = b_n$. (f need not be a surjection.)

(3) The process continues for ever.

Now for each $b \in B$ we have a well-defined process which can turn out in one of three ways, and so we can partition the set B into three mutually disjoint subsets. Let

B_A = all $b \in B$ such that the process ends with an a_n,

B_B = all $b \in B$ such that the process ends with a b_n,

and

B_∞ = all $b \in B$ such that the process never ends.

The same process can be applied starting with elements of A, and likewise A can be partitioned into three disjoint subsets. Let

A_A = all $a \in A$ such that the process ends with an a_n,

A_B = all $a \in A$ such that the process ends with a b_n,

and

$$A_\infty = \text{all } a \in A \text{ such that the process never ends.}$$

We require to show that $A \sim B$. We do this by demonstrating that $A_A \sim B_A$, $A_B \sim B_B$ and $A_\infty \sim B_\infty$. The restriction of f to A_A is a bijection from A_A to B_A. To prove this we must show two things:

    (a) $a \in A_A$ implies $f(a) \in B_A$,

and

    (b) for each $b \in B_A$ there is $a \in A_A$ with $f(a) = b$.

For (a), let $a \in A_A$. Then the process applied to a ends in A. Consider the process applied to $f(a)$. Its first step takes us back to a, and then it continues with the process applied to a, ending in A. Thus $f(a) \in B_A$, as required. For (b), let $b \in B_A$. Then the process applied to b ends in A, and in particular it must have a first stage (for otherwise it would end in B with b itself). Hence, $b = f(a)$ for some $a \in A$. But the process applied to this a is the same as the continuation of the process applied to b, and therefore it ends in A. Thus $a \in A_A$, as required, and we have shown that the restriction of f is a bijection from A_A to B_A.

By exactly the same argument we can show that $g : B_B \to A_B$ is a bijection, and consequently $g^{-1} : A_B \to B_B$ is a bijection. And lastly $f : A_\infty \to B_\infty$ is a bijection, for f is an injection and if $b \in B_\infty$ then $b = f(a)$ for some $a \in A$, since the process applied to b must start, and this a belongs to A_∞. This is because the process starting from a is the same as the process starting from b after the first step, and this never ends, since $b \in B_\infty$.

We can now define a bijection $F : A \to B$ by

$$F(x) = \begin{cases} f(x) & \text{if } x \in A_A \\ f(x) & \text{if } x \in A_\infty \\ g^{-1}(x) & \text{if } x \in A_B. \end{cases}$$

Verification that F is a bijection is left as an exercise, but follows from the facts that A_A, A_∞ and A_B are disjoint, B_A, B_∞ and B_B are disjoint, and f, f and g^{-1} are respectively bijections.

▶ The usefulness of this theorem will be demonstrated in numerous applications in the remainder of this chapter. Let us consider now the set $\mathbb{R}$ of real numbers, and intervals in $\mathbb{R}$. Some isolated results are given in the exercises preceding this section, for example $\mathbb{R} \sim \mathbb{R}^+$, $[0, 1] \sim [1, 3]$, $(-\pi/2, \pi/2) \sim \mathbb{R}$. We can now give a general result embracing all of these.

Theorem 2.19

Let I be any interval in $\mathbb{R}$ which is not empty and not a singleton. Then $I \sim \mathbb{R}$.

Proof

Consider first an open interval (a, b). This is equinumerous with the interval $(-\pi/2, \pi/2)$. The best way to see this is graphically, as shown in Fig. 2.2. The straight line is the graph of a bijection from

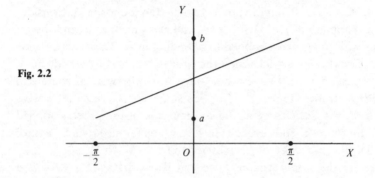

Fig. 2.2

$(-\pi/2, \pi/2)$ to (a, b). The line joins the points with coordinates $(-\pi/2, a)$ and $(\pi/2, b)$, so the positions of a and b on the y-axis do not affect the argument. This function is one-one since no two function values are equal and it is onto since every element of (a, b) corresponds to a point on the graph. The graph has equation $y = b + [(b - a)/\pi][x - (\pi/2)]$, $(-\pi/2 < x < \pi/2)$. Now since $(-\pi/2, \pi/2) \sim \mathbb{R}$ via the tangent function, we have $(a, b) \sim \mathbb{R}$, for every bounded non-empty open interval (a, b).

To complete the proof, let I be *any* interval in $\mathbb{R}$, not empty and not a singleton. Then I contains a bounded non-empty open interval, J say. We have $J \leqslant I$ necessarily, since $J \subseteq I$. Hence $\mathbb{R} \leqslant I$, by Theorem 2.5, since $\mathbb{R} \sim J$. But $I \subseteq \mathbb{R}$, so $I \leqslant \mathbb{R}$. By the Schröder–Bernstein theorem, then, we obtain $\mathbb{R} \sim I$, as required.

Corollary 2.20

Let I, J be any intervals in $\mathbb{R}$, which are not empty and not singletons. Then $I \sim J$.

Proof

$I \sim \mathbb{R}$ and $J \sim \mathbb{R}$, so $I \sim J$ by Theorem 2.1.

▶ These results will be useful to us in obtaining theorems about sets equinumerous with $\mathbb{R}$. But first let us use the Schröder–Bernstein theorem again to prove an important relationship between $\mathbb{R}$ and $\mathbb{N}$.

Theorem 2.21
$\mathbb{R} \sim P(\mathbb{N})$.

Proof
We show that $[0, 1) \sim P(\mathbb{N})$, and use the fact that $\mathbb{R} \sim [0, 1)$ to obtain the theorem. We require to construct injections $[0, 1) \to P(\mathbb{N})$ and $P(\mathbb{N}) \to [0, 1)$.

First define $f : P(\mathbb{N}) \to [0, 1)$ as follows. Given $X \subseteq \mathbb{N}$, we construct a decimal expansion $0 \cdot a_0 a_1 a_2 \cdots$ by putting

$$a_i = \begin{cases} 0 & \text{if } i \notin X \\ 1 & \text{if } i \in X. \end{cases}$$

We let $f(X) = 0 \cdot a_0 a_1 a_2 \ldots$. Certainly f is an injection, for if $f(X) = f(Y) = 0 \cdot a_0 a_1 a_2 \ldots$, then $i \in X \Leftrightarrow a_i = 1 \Leftrightarrow i \in Y$ and so $X = Y$. Hence, $P(\mathbb{N}) \preccurlyeq [0, 1)$.

Now we must define an injection $g : [0, 1) \to P(\mathbb{N})$. This is a little harder. First note that, by Theorem 1.29, an element of $[0, 1)$ can be expressed *uniquely* as a decimal

$$0 \cdot n_0 n_1 n_2 \ldots, \text{ with } 0 \le n_k \le 9,$$

provided that expressions ending with a repeating 9 are not permitted. Given $x \in [0, 1)$, write $x = 0 \cdot n_0 n_1 n_2 \cdots$ as above and let $g(x) = \{n_k 10^k : k \in \mathbb{N}\}$. Then $g : [0, 1) \to P(\mathbb{N})$ and g is an injection. For suppose that $g(x) = g(y)$, with $x = 0 \cdot m_0 m_1 m_2 \ldots$, and $y = 0 \cdot n_0 n_1 n_2 \ldots$. Let $k \in \mathbb{N}$. Now $m_k 10^k \in g(x)$, so $m_k 10^k \in g(y)$ also. Hence, $m_k 10^k = n_i 10^i$ for some $i \in \mathbb{N}$. Since m_k and n_i are single digit numbers, we must have $i = k$ and $m_k = n_k$. It follows that $x = y$. Hence, g is an injection $[0, 1) \to P(\mathbb{N})$, as required, and we conclude that $[0, 1) \sim P(\mathbb{N})$, using the Schröder–Bernstein theorem. It follows that $\mathbb{R} \sim P(\mathbb{N})$.

▶ By virtue of these new results we now know that there is a large collection of uncountable sets: $\mathbb{R}$, all intervals in $\mathbb{R}$, and $P(\mathbb{N})$. In fact, the ones we have found are all equinumerous. Sets which are equinumerous with $\mathbb{R}$ (and so with $P(\mathbb{N})$) have some properties analogous to those of countable sets.

Theorem 2.22

Let A, B be sets with $A \sim \mathbb{R}$ and $B \sim \mathbb{R}$. Then $A \cup B \sim \mathbb{R}$.

Proof

First suppose that $A \cap B = \emptyset$. There exist bijections $f : A \to [0, 1)$ and $g : B \to [1, 2)$, say. Combine these to obtain a bijection $h : A \cup B \to [0, 2)$ by

$$h(x) = \begin{cases} f(x) & \text{if } x \in A \\ g(x) & \text{if } x \in B. \end{cases}$$

Hence, $A \cup B \sim [0, 2)$, so $A \cup B \sim \mathbb{R}$.

If A and B are not disjoint then we can obtain at best an injection $k : A \cup B \to [0, 2)$ by

$$k(x) = \begin{cases} f(x) & \text{if } x \in A \\ g(x) & \text{if } x \in B \backslash A. \end{cases}$$

In general, k will not be a surjection since all of $[1, 2)$ will not be contained in the range of k. However, we obtain $A \cup B \leqslant [0, 2)$, so $A \cup B \leqslant \mathbb{R}$. But $\mathbb{R} \sim A$ and $A \leqslant A \cup B$, since $A \subseteq A \cup B$, so $\mathbb{R} \leqslant A \cup B$. Hence, $A \cup B \sim \mathbb{R}$ in this case also.

(The reader may note that it is not necessary in the above proof to treat the disjoint and non-disjoint cases separately. The second part of the proof covers the disjoint case also, but we have included both cases for the sake of clarity.)

Corollary 2.23

Let $A_1, A_2, \ldots, A_n$ be sets with $A_i \sim \mathbb{R}$ for $1 \leqslant i \leqslant n$. Then $A_1 \cup A_2 \cup \cdots \cup A_n \sim \mathbb{R}$.

Proof

By induction.

Theorem 2.24

Let A, B be sets with $A \sim \mathbb{R}$ and $B \sim \mathbb{R}$. Then $A \times B \sim \mathbb{R}$.

Proof

First consider $(0, 1) \times (0, 1)$. Define $f : (0, 1) \times (0, 1) \to (0, 1)$ by

$$f(x, y) = 0 \cdot a_0 b_0 a_1 b_1 \ldots,$$

where $x = 0 \cdot a_0 a_1 a_2 \ldots$, $y = 0 \cdot b_0 b_1 b_2 \cdots$ in decimal form. Then f is an injection, so $(0, 1) \times (0, 1) \leqslant (0, 1)$.

Now define $g:(0, 1) \to (0, 1) \times (0, 1)$ by

$$g(x) = (x, \tfrac{1}{2}).$$

Then g is an injection, so $(0, 1) \preccurlyeq (0, 1) \times (0, 1)$. Hence, by the Schröder–Bernstein theorem,

$$(0, 1) \times (0, 1) \sim (0, 1),$$

and so

$$(0, 1) \times (0, 1) \sim \mathbb{R}.$$

Suppose that $A \sim \mathbb{R}$ and $B \sim \mathbb{R}$. Then $A \sim (0, 1)$ and $B \sim (0, 1)$, say, via bijections $p:A \to (0, 1)$ and $q:B \to (0, 1)$. Define $r:A \times B \to (0, 1) \times (0, 1)$ by

$$r(a, b) = (p(a), q(b)).$$

It is an easy exercise to verify that r is a bijection, so $A \times B \sim (0, 1) \times (0, 1)$ and hence $A \times B \sim \mathbb{R}$.

Corollary 2.25
Let $A_1, A_2, \ldots, A_n$ be sets with $A_i \sim \mathbb{R}$ for $1 \le i \le n$. Then

$$A_1 \times A_2 \times \cdots \times A_n \sim \mathbb{R}.$$

Proof
A generalisation of the above procedure yields a proof. It is left as an exercise.

It is worth noting two special cases of this result.

Corollary 2.26
(i) $\mathbb{R}^n \sim \mathbb{R}$. The points of n-dimensional Euclidean space form a set equinumerous with $\mathbb{R}$.

(ii) $\mathbb{C} \sim \mathbb{R}$.

See Example 2.3(b).

▶ We can treat also the case of a union of a countable collection of sets equinumerous with $\mathbb{R}$. Compare the proof of the following with that of Theorem 2.14, and note that the axiom of choice is implicitly used.

Theorem 2.27
Let $A_0, A_1, A_2, \ldots$ be sets with $A_i \sim \mathbb{R}$ for $i \in \mathbb{N}$. Then $\bigcup_{i \in \mathbb{N}} A_i \sim \mathbb{R}$.

Proof
There exist bijections $f_i : A_i \to [i, i+1)$ $(i \in \mathbb{N})$, so let us construct
a function $F : \bigcup_{i \in \mathbb{N}} A_i \to \mathbb{R}$ by

$$F(x) = \begin{cases} f_0(x) & \text{if } x \in A_0, \\ f_1(x) & \text{if } x \in A_1 \backslash A_0, \\ f_2(x) & \text{if } x \in A_2 \backslash (A_0 \cup A_1), \\ \text{etc.} \end{cases}$$

F is easily seen to be an injection, so $\bigcup_{i \in \mathbb{N}} A_i \preccurlyeq \mathbb{R}$. Also, $A_0 \subseteq \bigcup_{i \in \mathbb{N}} A_i$
and $A_0 \sim \mathbb{R}$, so $\mathbb{R} \preccurlyeq \bigcup_{i \in \mathbb{N}} A_i$. By the Schröder–Bernstein theorem, then,
$\bigcup_{i \in \mathbb{N}} A_i \sim \mathbb{R}$.

Example 2.28
The set of all finite subsets of $\mathbb{R}$ is equinumerous with $\mathbb{R}$.

A proof of this is a useful exercise on the methods of this chapter.
First fix $n \in \mathbb{Z}^+$, and denote by $\boldsymbol{P}_n(\mathbb{R})$ the set of all n-element subsets of
$\mathbb{R}$. We show that $\boldsymbol{P}_n(\mathbb{R}) \sim \mathbb{R}$. Given $X \in \boldsymbol{P}_n(\mathbb{R})$, order the elements of X
by magnitude to obtain an ordered n-tuple, an element of $\mathbb{R}^n$. This yields
an injection from $\boldsymbol{P}_n(\mathbb{R})$ to $\mathbb{R}^n$, so $\boldsymbol{P}_n(\mathbb{R}) \preccurlyeq \mathbb{R}^n$, and hence $\boldsymbol{P}_n(\mathbb{R}) \preccurlyeq \mathbb{R}$. Now,
given $x \in \mathbb{R}$, we can construct the set $\{x, x+1, \ldots, x+n-1\} \in \boldsymbol{P}_n(\mathbb{R})$, and
this rule yields an injection from $\mathbb{R}$ to $\boldsymbol{P}_n(\mathbb{R})$. Hence, $\mathbb{R} \preccurlyeq \boldsymbol{P}_n(\mathbb{R})$. By the
Schröder–Bernstein theorem, then, $\boldsymbol{P}_n(\mathbb{R}) \sim \mathbb{R}$, and this holds for every
$n \in \mathbb{Z}^+$. The set of all finite subsets of $\mathbb{R}$ is just the union of all the sets
$\boldsymbol{P}_n(\mathbb{R})$ for $n \in \mathbb{Z}^+$. The required result is therefore a consequence of
Theorem 2.27.

Example 2.29
The set of all irrational numbers is equinumerous with $\mathbb{R}$.

This is a little bit tricky. First we demonstrate a preliminary lemma:
if $A \sim B$ and $A \cap C = B \cap C = \emptyset$ then $A \cup C \sim B \cup C$. To see this let
$f : A \to B$ be a bijection. Then $g : A \cup C \to B \cup C$ is a bijection, where

$$g(x) = \begin{cases} f(x) & \text{if } x \in A \\ x & \text{if } x \in C. \end{cases}$$

Verification is left as an exercise.

Now let the set of irrational numbers be denoted by X, and let P be
some fixed countable proper subset of X, say

$$P = \{\sqrt{p} : p \text{ is a prime number}\}.$$

If we write $Y = X \backslash P$, then

$$\mathbb{R} = \mathbb{Q} \cup X = \mathbb{Q} \cup (P \cup Y) = (\mathbb{Q} \cup P) \cup Y.$$

Now $\mathbb{Q} \cup P$ is the union of two countable sets, so is countable, and is clearly infinite. Hence, $\mathbb{Q} \cup P \sim P$. By the above lemma then, we have

$$(\mathbb{Q} \cup P) \cup Y \sim P \cup Y$$

i.e.

$$\mathbb{R} \sim X.$$

▶ Sets which are neither countable nor equinumerous with $\mathbb{R}$ do exist. For example $P(\mathbb{R})$, the set of all subsets of $\mathbb{R}$, is such a set by virtue of Theorem 2.16. The set of all subsets of $P(\mathbb{R})$ is another such, and clearly this process can be continued indefinitely, so there is no bound on the 'size' of a set. However, when such large sets are under consideration, intuitive ideas about sizes of sets become less applicable, and more formal mathematical treatment is necessary. The study of large cardinal numbers is an important area of the foundations of mathematics and it is one where the underlying principles are still not generally agreed.

One very basic aspect of this is worth discussing at this stage. Our considerations have led to the discovery of infinite sets equinumerous with $\mathbb{N}$ and, at the next level up, infinite sets equinumerous with $\mathbb{R}$. Are there any sets 'in between'? To phrase the question more specifically, are there any subsets of $\mathbb{R}$ which are neither countable nor equinumerous with $\mathbb{R}$? We are certainly not in a position to answer this question immediately, since all of the subsets of $\mathbb{R}$ which we have come across have fallen into one of these two categories. The conjecture that there are no sets with size 'in between' $\mathbb{N}$ and $\mathbb{R}$ was originally made by Cantor towards the end of the nineteenth century, and is known as the *continuum hypothesis*. Intuition gives little guidance, for although $\mathbb{R}$ is apparently a much 'bigger' set than $\mathbb{N}$, all familiar ways of describing or constructing subsets of $\mathbb{R}$ lead either to countable sets or to sets equinumerous with $\mathbb{R}$. The more formal methods described later in this book have been brought to bear also on this problem, with surprising consequences (see Theorem 6.40). It has been shown (Gödel, 1938) that the continuum hypothesis is consistent with the standard axioms for set theory, i.e. that no contradiction would follow if it were taken to be true. However, it has also been shown (Cohen, 1963) that it is independent of the other standard axioms of set theory, i.e. it cannot be derived as a consequence of those axioms, and hence that it is consistent to suppose that the continuum hypothesis is false. This brings out an inadequacy in the standard axioms for set theory, since they do not demonstrate the truth or falsity of such a simple principle. However, there is a deeper inadequacy which causes this, and that is in our intuitive understanding of infinite sets and their properties, for the axioms merely reflect properties which are intuitively clear. There are thus fundamental mathematical problems at a very elementary level in this part of the subject.

Exercises

1. Let A, B, C be sets such that $A \subseteq B \subseteq C$ and $A \sim C$. Prove that $A \sim B$ and $B \sim C$. Suppose that $A_1 \leqslant A_2 \leqslant \cdots \leqslant A_n$ and $A_1 \sim A_n$. Prove that $A_1 \sim A_2 \sim \cdots \sim A_n$.

2. Let A and B be sets with $A \sim B$ and $A \cap B = \emptyset$. Show that $A \cup B \sim A \times \{0, 1\}$.

3. Prove that if $A \leqslant B$ then $P(A) \leqslant P(B)$. Deduce that if $A \sim B$ then $P(A) \sim P(B)$.

4. Given sets A, B, C and D such that $A \leqslant C$ and $B \leqslant D$, prove that $A \times B \leqslant C \times D$.

5. Let X be a set satisfying $X \times X \sim X$. Prove that, for any set Y, if $Y \leqslant X$ then $X \cup Y \sim X$.

6. Prove the following:
   (i) If A is finite and $B \sim \mathbb{R}$ then $A \cup B \sim \mathbb{R}$ and $A \times B \sim \mathbb{R}$.
   (ii) If A is countable and $B \sim \mathbb{R}$ then $A \cup B \sim \mathbb{R}$ and $A \times B \sim \mathbb{R}$.

7. Show that the set of all finite sequences of elements of $\mathbb{R}$ is equinumerous with $\mathbb{R}$.

8. (Harder) Show that the set of all infinite sequences of elements of $\mathbb{N}$ is equinumerous with $\mathbb{R}$. (Hint: consider a sequence as a set $\{(n, a_n) : n \in \mathbb{N}\}$ of ordered pairs.)

9. Show that the set of all straight lines in a plane is equinumerous with $\mathbb{R}$. The line with equation $3y = 2x - 5$ has the property that whenever x is rational, the corresponding point on the line has a rational y-coordinate. What can be said about the size of the set of all such lines?

10. Prove that every subset of $\mathbb{R}$ which contains a non-empty open interval is equinumerous with $\mathbb{R}$. Give an example of a subset of $\mathbb{R}$ which is equinumerous with $\mathbb{R}$ but which contains no non-empty open interval.

11. Prove that the set of all infinite subsets of $\mathbb{N}$ is equinumerous with $P(\mathbb{N})$. (See Example 2.29 for similar proof.)

12. Prove the following:
    (i) $\{X \in P(\mathbb{N}) : \mathbb{N} \backslash X \text{ is finite}\} \sim \mathbb{N}$.
    (ii) $\{X \in P(\mathbb{N}) : \mathbb{N} \backslash X \text{ is infinite}\} \sim P(\mathbb{N})$.

13. (Harder) Prove that the set of all infinite sequences of elements of $\mathbb{R}$ is equinumerous with $\mathbb{R}$. (Hint: Given an infinite sequence of real numbers, express each in decimal form and write the sequence vertically so that an array is obtained. By a diagonal listing procedure, as given

after Theorem 2.14, this array gives rise to a single infinite sequence of natural numbers. In this way we can construct an injection from the given set into the set referred to in Exercise 8 above.)

2.3 Cardinal numbers

The cardinal number of a set will be a measure of its size, in the sense adopted in the first section of this chapter. The definition of exactly what sort of object a cardinal number is will have to wait until Chapter 6, since cardinal number is a difficult notion. For the moment we shall regard it as a convenient form of words for describing a familiar situation.

Definition

Sets A and B *have the same cardinal number* if there is a bijection between them.

Implicit in this is the idea that a cardinal number is something that equinumerous sets have in common. For finite sets there is no difficulty about this; here the cardinal number may be taken as the number of elements in the set. For infinite sets, we have already found certain categories of sets, the principal examples being those equinumerous with $\mathbb{N}$ and those equinumerous with $\mathbb{R}$. We shall introduce symbols to denote the two corresponding infinite cardinal numbers as follows:

$\aleph_0$ (aleph nought) is the cardinal number of $\mathbb{N}$.

$\aleph$ (aleph) is the cardinal number of $\mathbb{R}$.

For the moment, however, saying that a set 'has cardinal number $\aleph_0$' means no more than that it is equinumerous with $\mathbb{N}$ (and similarly for $\aleph$ in relation to $\mathbb{R}$).

Notation

We shall use lower case Greek letters κ, λ, μ, ... to denote cardinal numbers, and for sets A and B we shall use the abbreviations card A = card B and card $A = \kappa$.

Remark

It is important to note that all statements and results about cardinal numbers in this chapter are statements about bijections between sets, or, as in the next definition, about injections.

Definition

We say that card $A \leqslant$ card B if A is dominated by B, i.e. if there is an injection from A to B.

Many of our earlier results can be re-stated in these terms, for example the following.

(Theorem 2.1)   If card $A =$ card B and card $B =$ card C, then card $A =$ card C.

(Theorem 2.4)   If $A \subseteq B$, then card $A \leqslant$ card B.

(Theorem 2.15) Card $\mathbb{Q} = \aleph_0$.

(Theorem 2.16) For any set A, card $A \neq$ card $\boldsymbol{P}(A)$.

(Theorem 2.18) If card $A \leqslant$ card B and card $B \leqslant$ card A, then card $A =$ card B.

(Theorem 2.21) Card $\boldsymbol{P}(\mathbb{N}) = \aleph$.

▶ So far in this section we have merely dressed up old ideas in new terminology. We shall continue to do this, but new ideas will enter into it along the way. We shall find that some familiar operations on sets, for example union and Cartesian product, translate into well-defined operations on cardinal numbers. Let us consider finite sets first of all.

Given sets A and B with, respectively, m elements and n elements, we have the following:

If $A \cap B = \emptyset$ then $A \cup B$ contains $m + n$ elements.
$A \times B$ contains $m \times n$ elements (whether $A \cap B = \emptyset$ or not).
$\boldsymbol{P}(A)$ contains 2^m elements.

If these results are not familiar they may be treated as exercises.

Similar results hold for infinite sets when we extend the notions of sum, product and powers.

Theorem 2.30

Let A, B, C and D be sets with $A \sim C$ and $B \sim D$. Then
(i) If $A \cap B = \emptyset$ and $C \cap D = \emptyset$ then $A \cup B \sim C \cup D$.
(ii) $A \times B \sim C \times D$.

Proof

Let $f : A \to C$ and $g : B \to D$ be bijections.
(i) $h : A \cup B \to C \cup D$ is a bijection, where

$$h(x) = \begin{cases} f(x) & \text{if } x \in A \\ g(x) & \text{if } x \in B. \end{cases}$$

(ii) $k : A \times B \to C \times D$ is a bijection, where

$$k(x, y) = (f(x), g(y)) \quad (x \in A, y \in B).$$

In both cases, verification is left as an exercise.

► This theorem enables us to make the following definition.

Definition
Let κ and λ be any two cardinal numbers.
  (i) The *sum*, $\kappa + \lambda$, is the cardinal number of $A \cup B$, where A and B are any sets with card $A = \kappa$, card $B = \lambda$ and $A \cap B = \emptyset$.
  (ii) The *product*, $\kappa\lambda$, is the cardinal number of $A \times B$, where A and B are any sets with card $A = \kappa$ and card $B = \lambda$.

Note that Theorem 2.30 says precisely that these notions are well-defined, that is that the choice of the sets A and B does not affect the resulting sum and product of κ and λ.

For finite cardinal numbers, sums and products under this definition will certainly agree with the normal operations on natural numbers. When infinite cardinal numbers are involved we obtain results which appear completely different.

Theorem 2.31
  (i) $n + \aleph_0 = \aleph_0, \quad n\aleph_0 = \aleph_0 \quad (n \in \mathbb{Z}^+)$.
  (ii) $\aleph_0 + \aleph_0 = \aleph_0, \quad \aleph_0\aleph_0 = \aleph_0$.
  (iii) $n + \aleph = \aleph, \quad n\aleph = \aleph \quad (n \in \mathbb{Z}^+)$.
  (iv) $\aleph_0 + \aleph = \aleph, \quad \aleph_0\aleph = \aleph$.
  (v) $\aleph + \aleph = \aleph, \quad \aleph\aleph = \aleph$.

Proof
  (i) Let card $A = n$, card $B = \aleph_0$, and $A \cap B = \emptyset$. Then $A \cup B$ is countable (Theorem 2.9) and clearly infinite since B is infinite. Hence, $A \cup B \sim \mathbb{N}$, and so card$(A \cup B) = \aleph_0$. This yields $n + \aleph_0 = \aleph_0$. Similarly, $A \times B \sim \mathbb{N}$ (using Theorem 2.12) and consequently $n\aleph_0 = \aleph_0$.

  (ii) Essentially the same proof as for (i).

  (iii) Let $A = \{1, 2, \ldots, n\}$ and $B = (n, n+1)$ (interval in $\mathbb{R}$), say, so that card $A = n$ and card $B = \aleph$. Then $A \cup B \subseteq \mathbb{R}$, so card $A \cup B \leqslant \aleph$. Also, $B \subseteq A \cup B$, so $\aleph \leqslant$ card $A \cup B$. Hence, by the Schröder–Bernstein theorem, we have card $A \cup B = \aleph$, i.e.

$n + \aleph = \aleph$. Further,

$$A \times B = \{(x, y) : x \in A \ \& \ y \in B\}$$
$$= \{(1, y) : y \in B\} \cup \{(2, y) : y \in B\} \cup \cdots \cup \{(n, y) : y \in B\}$$
$$= \text{a union of finitely many sets each with cardinal}$$
$$\text{number } \aleph.$$

By Corollary 2.33, then, card $A \times B = \aleph$, and so $n\aleph = \aleph$.

(iv) Let $A = \mathbb{N}$, $B = (-1, 0)$ (interval in $\mathbb{R}$). Then card $A \cup B = \aleph$ (proof just as in (iii) above), so $\aleph_0 + \aleph = \aleph$. Further,

$$A \times B = \{(0, y) : y \in B\} \cup \{(1, y) : y \in B\} \cup \cdots$$
$$= \text{a union of a countable collection of sets with cardinal}$$
$$\text{number } \aleph.$$

By Theorem 2.27, then, card $A \times B = \aleph$, and so $\aleph_0\aleph = \aleph$.

(v) These follow immediately from Theorems 2.22 and 2.24.

▶ These are some examples of the sum and product operation on cardinal numbers. Consideration of other cardinal numbers involves exponentiation (which concerns the power set operation), but before investigating that let us note some properties of the sum and product in general.

Theorem 2.32
Let κ, λ, μ be cardinal numbers.
(i) $\kappa + \lambda = \lambda + \kappa$, $\quad \kappa\lambda = \lambda\kappa$.
(ii) $(\kappa + \lambda) + \mu = \kappa + (\lambda + \mu)$, $\quad (\kappa\lambda)\mu = \kappa(\lambda\mu)$.
(iii) $\kappa(\lambda + \mu) = \kappa\lambda + \kappa\mu$.

Proof
These are easy exercises. We give two of the proofs and the reader can supply the others.

(i) Let $\kappa = $ card A, $\lambda = $ card B, with $A \cap B = \emptyset$. Then $\kappa + \lambda = $ card$(A \cup B)$, and $\lambda + \kappa = $ card$(B \cup A)$, and certainly we have card$(A \cup B) = $ card$(B \cup A)$.

(ii) Let $\kappa = $ card A, $\lambda = $ card B, $\mu = $ card C, with $B \cap C = \emptyset$. Then $\kappa(\lambda + \mu) = $ card$(A \times (B \cup C))$, $\kappa\lambda = $ card$(A \times B)$, and $\kappa\mu = $ card$(A \times C)$. Now $(A \times B) \cap (A \times C) = \emptyset$, since $B \cap C = \emptyset$, and so $\kappa\lambda + \kappa\mu = \text{card}((A \times B) \cup (A \times C))$. But

$$A \times (B \cup C) = (A \times B) \cup (A \times C),$$

so $\kappa(\lambda + \mu) = \kappa\lambda + \kappa\mu$, as required.

▶ Now let us consider exponentiation. For a finite set A with n elements, we have seen that $P(A)$ contains 2^n elements. We shall make an appropriate definition of exponentiation of cardinal numbers so that this result extends to infinite sets also, i.e. if card $A = \kappa$ then card $P(A) = 2^\kappa$ (the notation has still to be explained).

Consider an arbitrary set A. For each subset X of A we can define a function C_X from A to $\{0, 1\}$ as follows:

$$C_X(y) = \begin{cases} 1 & \text{if } y \in X \\ 0 & \text{if } y \in A\backslash X. \end{cases}$$

C_X is called the *characteristic function* of X. We thus have a correspondence between subsets of A and functions from A to $\{0, 1\}$. This correspondence is in fact a bijection. Before verifying this, let us introduce some notation.

Notation

For any sets A and B, the set of all functions from A to B is denoted by B^A. Note that the domain set is the 'exponent'.

In our discussion above, the set $\{0, 1\}^A$ has occurred, indeed we have found a function, say F, from $P(A)$ to $\{0, 1\}^A$, given by $F(X) = C_X$. This F is an injection, for suppose that $X \subseteq A$, $Z \subseteq A$ and $C_X = C_Z$. Then $C_X(y) = C_Z(y)$ for every $y \in A$, and hence $C_X(y) = 1$ if and only if $C_Z(y) = 1$ (for $y \in A$). It follows that for any given $y \in A$, $y \in X$ if and only if $y \in Z$, so $X = Z$. Also, F is a surjection, for if ϕ is a function from A to $\{0, 1\}$ then the set $\{y \in A : \phi(y) = 1\}$ (call it Y) is a subset of A and $F(Y) = \phi$. We have therefore proved the following.

Theorem 2.33
For any set A, $P(A) \sim \{0, 1\}^A$.

▶ This result will translate very nicely into a result about cardinal numbers, but first we give the definition of exponentiation.

Definition

Let κ and λ be any cardinal numbers. Then λ^κ is the cardinal number of the set B^A, where A and B are any sets such that card $A = \kappa$ and card $B = \lambda$.

Of course this definition, like the previous one, requires a theorem to ensure that it makes sense; that λ^κ does not depend on the particular choice of sets A and B.

Theorem 2.34

Let A, B, C and D be sets with $A \sim C$ and $B \sim D$. Then $B^A \sim D^C$.

Proof

An exercise for the reader.

We now have

Theorem 2.35

For any set A, with cardinal number κ, the cardinal number of $P(A)$ is 2^κ (the cardinal number of $\{0, 1\}$ is 2).

▶ Apart from being an elegant analogue of the result for finite sets, this theorem (along with Theorem 2.21) enables us to relate our two familiar infinite cardinal numbers $\aleph_0$ and $\aleph$.

Corollary 2.36

$\aleph = 2^{\aleph_0}$.

Proof

$P(\mathbb{N})$ has cardinal number $2^{\aleph_0}$, by Theorem 2.35. But $P(\mathbb{N}) \sim \mathbb{R}$, by Theorem 2.21, so $\mathbb{R}$ has cardinal number $2^{\aleph_0}$. Hence $\aleph = 2^{\aleph_0}$.

▶ This notation also gives a convenient way of writing down an unending sequence of ever larger infinite cardinal numbers, namely,

$$\aleph_0, 2^{\aleph_0}, 2^{2^{\aleph_0}}, 2^{2^{2^{\aleph_0}}} \ldots,$$

where each is the cardinal number of the power set of a set whose cardinal number is the preceding member of the sequence, and so, by Theorem 2.16, is strictly larger.

Let us now complete this section with some particular results concerning exponentiation.

Theorem 2.37

(i) $\aleph_0^n = \aleph_0$ $(n \in \mathbb{Z}^+)$.

(ii) $n^{\aleph_0} = \aleph$ $(n \in \mathbb{Z}^+)$.

(iii) $\aleph_0^{\aleph_0} = \aleph$.

(iv) $\aleph^n = \aleph$ $(n \in \mathbb{Z}^+)$.

(v) $\aleph^{\aleph_0} = \aleph$.

Proof

(i) Let $A = \{1, 2, \ldots, n\}$. We require to show that $\mathbb{N}^A$, the set of all functions from A to $\mathbb{N}$, is countable. But $\mathbb{N}^A \sim \mathbb{N} \times \mathbb{N} \times \cdots \times \mathbb{N}$ with n factors in the Cartesian product. To see this we can define a bijection ϕ as follows:

given $f : A \to \mathbb{N}$, let $\phi(f) = (f(1), f(2), \ldots, f(n))$

(verification is easy – we are merely associating a function f with the ordered n-tuple of its values). We know from Theorem 2.13, however, that $\mathbb{N} \times \mathbb{N} \times \cdots \times \mathbb{N}$ is countable. Hence $\mathbb{N}^A$ is countable. It is clearly infinite, so card $\mathbb{N}^A = \aleph_0$, and thus $\aleph_0^n = \aleph_0$.

(ii) We need a new technique here. Let $A = \{1, 2, \ldots, n\}$ as above. $A^{\mathbb{N}}$ is the set of all functions from $\mathbb{N}$ to A. Now a function f from $\mathbb{N}$ to A may be considered to be a set of ordered pairs (x, y) with $x \in \mathbb{N}$ and $y \in A$, with $y = f(x)$. This set is sometimes called the *graph* of the function. Thus if $f \in A^{\mathbb{N}}$ then $f \subseteq \mathbb{N} \times A$, i.e. $f \in P(\mathbb{N} \times A)$. In other words, $A^{\mathbb{N}} \subseteq P(\mathbb{N} \times A)$. Now $\mathbb{N} \times A$ is countable (Theorem 2.12) and infinite, so $\mathbb{N} \times A \sim \mathbb{N}$, and so $P(\mathbb{N} \times A) \sim P(\mathbb{N})$. By Theorem 2.4, we have $A^{\mathbb{N}} \leqslant P(\mathbb{N} \times A)$ so $A^{\mathbb{N}} \leqslant P(\mathbb{N})$, and hence card $A^{\mathbb{N}} \leqslant$ card $P(\mathbb{N})$, i.e. $n^{\aleph_0} \leqslant 2^{\aleph_0}$.

To complete this proof we must show that $2^{\aleph_0} \leqslant n^{\aleph_0}$. But $\{1, 2\} \subseteq \{1, 2, \ldots, n\}$, so $\{1, 2\}^{\mathbb{N}} \subseteq A^{\mathbb{N}}$, so card $\{1, 2\}^{\mathbb{N}} \leqslant$ card $A^{\mathbb{N}}$, giving the required result. The Schröder–Bernstein theorem (in its cardinal number formulation) then yields $n^{\aleph_0} = 2^{\aleph_0}$, and we know that $2^{\aleph_0} = \aleph$.

(iii) The technique of proof is identical with that used in the proof of (ii) above, so this is left as an exercise.

(iv) The method here is similar to part (i), using Corollary 2.36.

(v) $\aleph^{\aleph_0}$ is the cardinal number of $\mathbb{R}^{\mathbb{N}}$, i.e. the set of all functions from $\mathbb{N}$ to $\mathbb{R}$. A function f from $\mathbb{N}$ to $\mathbb{R}$ may be considered as an infinite sequence $f(0), f(1), \ldots$ of elements of $\mathbb{R}$, so we can think of $\mathbb{R}^{\mathbb{N}}$ as the set of all infinite sequences of real numbers. The reader is now left to complete the proof. A further hint is given in Exercise 13 on page 72.

▶ It can be shown, by a difficult argument using the axiom of choice, that for every infinite cardinal number κ, $\kappa\kappa = \kappa$ (see Corollary 5.16). A consequence of this is the following surprising theorem, which says that addition and multiplication of *infinite* cardinal numbers is rather trivial.

Theorem 2.38
For any infinite cardinal numbers κ and λ, if $\kappa \leq \lambda$ then

(i) $\kappa + \lambda = \lambda$,

and

(ii) $\kappa\lambda = \lambda$.

Proof
We use the results of Exercise 6 immediately following. Let $\kappa \leq \lambda$. Then

$$\lambda \leq \kappa + \lambda \leq \lambda + \lambda = 2\lambda \leq \lambda\lambda = \lambda,$$

and, consequently, $\kappa + \lambda = \lambda$. Also,

$$\lambda \leq \kappa\lambda \leq \lambda\lambda = \lambda,$$

and so $\kappa\lambda = \lambda$.

▶ A comprehensive reference for further results about the arithmetic of cardinal numbers is the book by Sierpinski.

This chapter has been based on informal intuitive ideas of sets and numbers, sufficient to grasp the concepts of countable and uncountable sets. Some important questions have been swept under the carpet, however. One of these is the nature of cardinal numbers. Another is the question of whether every set has a cardinal number (we can answer this question only when we know precisely what a cardinal number is). Another is the following: given any two cardinal numbers κ and λ, is it necessarily the case that either $\kappa \leq \lambda$ or $\lambda \leq \kappa$? Equivalent to this is the question: given any two sets A and B, is it necessarily the case that either $A \leq B$ or $B \leq A$? The continuum hypothesis is another difficulty we have already commented on. Problems such as these require a deeper analysis of the underlying principles of the subject, which we shall explore in later chapters.

Exercises 2.3

1. Prove that there is no infinite set A such that card $A \leq \aleph_0$ and card $A \neq \aleph_0$.

2. Is it the case that, for every infinite set A, we have $\aleph_0 \leqslant \mathrm{card}\, A$?

3. Prove the following, for any cardinal numbers κ, λ, μ.
   (i) $\kappa\lambda = \lambda\kappa$.
   (ii) $(\kappa + \lambda) + \mu = \kappa + (\lambda + \mu)$.
   (iii) $(\kappa\lambda)\mu = \kappa(\lambda\mu)$.

4. Prove the following, for every cardinal number κ.
   (i) $\kappa + \kappa = 2\kappa$.
   (ii) $\kappa\kappa = \kappa^2$.
   (iii) $\kappa + 0 = \kappa$.
   (iv) $\kappa 0 = 0$.
   (0 is the cardinal number of the empty set.)

5. Show that cancellation is invalid in equations involving sums and products of cardinal numbers, i.e. find counterexamples which demonstrate the following.
   (i) $\kappa + \mu = \lambda + \mu$ does not imply $\kappa = \lambda$.
   (ii) $\kappa\mu = \lambda\mu$ does not imply $\kappa = \lambda$.

6. Let κ, λ and μ be cardinal numbers with $\kappa \leqslant \lambda$. Prove that $\kappa + \mu \leqslant \lambda + \mu$ and $\kappa\mu \leqslant \lambda\mu$.

7. Let κ and λ be cardinal numbers with $\kappa \leqslant \lambda$. Show that there is a cardinal number μ such that $\lambda = \kappa + \mu$.

8. Let κ be a cardinal number with $\aleph_0 \leqslant \kappa$. Prove that $\aleph_0 + \kappa = \kappa$.

9. Prove, for any sets A, B and C, that if $A \subseteq B$ then $A^C \subseteq B^C$. Can we deduce also that $C^A \subseteq C^B$?

10. Let A, B, C and D be sets with $A \sim C$ and $B \sim D$. Show that $B^A \sim D^C$. (Theorem 2.34).

11. Prove that $\aleph_0^{\aleph_0} = \aleph$.

12. Prove that $\aleph^n = \aleph$ for any $n \in \mathbb{Z}^+$.

13. Prove or disprove the following.
    (i) For any cardinal numbers κ, λ and μ, $\kappa^{\lambda + \mu} = \kappa^\lambda \kappa^\mu$.
    (ii) For any cardinal numbers κ, λ and μ, $(\kappa^\lambda)^\mu = \kappa^{\lambda\mu}$.
    (iii) If $\kappa^\mu = \lambda^\mu$ then $\kappa = \lambda$.
    (iv) If $\mu^\kappa = \mu^\lambda$ then $\kappa = \lambda$.

14. Verify the following, where $n \in \mathbb{N}$, $n > 2$.

$$\aleph = 2^{\aleph_0} \leqslant n^{\aleph_0} \leqslant \aleph_0^{\aleph_0} \leqslant \aleph^{\aleph_0} = (2^{\aleph_0})^{\aleph_0} = 2^{\aleph_0 \aleph_0} = 2^{\aleph_0} = \aleph.$$

Consequently the $\leqslant$ signs may all be replaced by $=$ signs.

Further reading

Sierpinski [22] The standard reference work, representing the complete state of knowledge at the time that it was written (1952).

Stewart & Tall [23] A straightforward and easy to read introduction to the foundations of mathematics.

Swicrczkowski [25] A useful little book, whose content is limited to the ideas of this chapter.

3

ORDERED SETS

Summary

Starting with the general abstract definition of a relation, the various sorts of order relations are described and defined, illustrated by many examples. The notion of order isomorphism is introduced. Lattices and Boolean algebras are defined. Examples of these are given and some simple properties derived. Section 3.2 is not a prerequisite for later chapters, although there is reference in Chapter 5 to some of its results.

The reader is presumed to have some experience with abstract algebraic ideas. There is no dependence on the results in Chapters 1 and 2, but in some examples ideas from these chapters are used.

3.1 **Order relations and ordered sets**

The notion of a relation is fundamental in mathematics. Like the notion of a set, it is extremely general and consequently it crops up everywhere. We shall start from the beginning, with the broadest definition, but before we do that, let us observe that there are three kinds of relation which are particularly important, namely, functions, equivalence relations and order relations. Every student of mathematics should know what a function is and how central is the role played by functions in all branches of mathematics. Also, equivalence relations should be familiar although they are perhaps less pervasive. The idea of an order relation, as a general notion, is perhaps less well known, since much of mathematics needs to refer only to particular order relations and does not need to use the general notion or its properties. The most familiar examples are the standard orderings of the number systems (by magnitude) and the ordering of a collection of sets by inclusion.

Definitions

A *relation* from a set X to a set Y is a subset of the Cartesian product $X \times Y$. A *relation on a set* X is a subset of $X \times X$. If R is such a relation and $(x, y) \in R$, we say that x is related to y, and for convenience we may write xRy.

This is the definition of a *binary* relation. It can clearly be extended to the case of an *n-ary relation*, i.e. a subset of a Cartesian product $X_1 \times X_2 \times \cdots \times X_n$.

A *function* is a relation which is single valued. A subset, R, of $X \times Y$, is a function from X to Y if for each $x \in X$ there is precisely one $y \in Y$ with $(x, y) \in R$. Notice that we are identifying a function with its graph – a function is to be a set of ordered pairs, not a rule for calculating values.

A function can have more than one argument, of course. An n-place function is a function whose domain set (X in the above definition) is a Cartesian product of n sets.

An *equivalence relation* on a set X is a binary relation from X to X (i.e. a subset of $X \times X$) which is reflexive, symmetric and transitive.

This is assumed to be a familiar notion, but it will do no harm to refresh our memories about these words. For a binary relation R on a set X:

R is *reflexive* if xRx for every $x \in X$.

R is *symmetric* if xRy implies yRx for every $x, y \in X$.

R is *transitive* if xRy and yRz implies xRz for every $x, y, z \in X$.

The significance of the equivalence relation lies in the resultant partition of the underlying set into disjoint equivalence classes. This idea is used widely, particularly in algebra, and indeed has already been used in this book in the construction of the number systems.

All of these definitions will be well known to most readers, so let us now proceed to the business of the chapter.

Definition

A binary relation on a set X is an *order* relation if it is

(i) reflexive, i.e. xRx for every $x \in X$,

(ii) anti-symmetric, i.e. xRy and yRx imply $x = y$, for every $x, y \in X$,

and

(iii) transitive, i.e. xRy and yRz imply xRz, for every $x, y, z \in X$.

We say that X is *ordered* by R.

Examples 3.1

(a) The standard order ($\leqslant$) by magnitude on the set $\mathbb{N}$ is an order relation if we make it fit the abstract definition as follows. Let

R be the set of all ordered pairs (m, n) where $m, n \in \mathbb{N}$ and $m \leqslant n$. It is easy then to verify that R is a binary relation on $\mathbb{N}$ which is reflexive, anti-symmetric and transitive.

(b) The standard orders by magnitude on $\mathbb{Z}$ and on $\mathbb{Q}$ and on $\mathbb{R}$ are order relations as above.

(c) The relation $\subseteq$ on a collection of sets is an order relation. Let A be a set whose elements are sets. Then let

$$I_A = \{(X, Y) \in A \times A : X \subseteq Y\}.$$

Then I_A is an order relation on A.

(d) The relation 'divides' on the set $\mathbb{Z}^+$ is an order relation. Verification is left to the reader.

(e) The relation R on the set $\mathbb{Z} \times \mathbb{Z}$ defined as follows is an order relation.

$(a, b)R(x, y)$ if and only if $a \leqslant x$ and $b \leqslant y$.

A simple theorem might help to consolidate these ideas. Let R be any binary relation. By the *converse* of R we mean the set $\{(y, x) : (x, y) \in R\}$. This is denoted by R^{-1}.

Theorem 3.2
The converse of an order relation is an order relation.

Proof

Let R be an order relation on the set X. $(x, x) \in R$, so $(x, x) \in R^{-1}$, for each $x \in X$, so R^{-1} is reflexive. Let $(x, y) \in R^{-1}$ and $(y, x) \in R^{-1}$. Then $(y, x) \in R$ and $(x, y) \in R$, and consequently $x = y$ since R is anti-symmetric. Thus R^{-1} is anti-symmetric. Lastly, suppose that $(x, y) \in R^{-1}$ and $(y, z) \in R^{-1}$. Then $(y, x) \in R$ and $(z, y) \in R$. Since R is transitive, we have $(z, x) \in R$, and hence $(x, z) \in R^{-1}$. R^{-1} is therefore transitive and the proof is complete.

▶ Restricting an order relation yields an order relation. Let us make this more precise. Let R be any binary relation on a set X and let $Y \subseteq X$. The *restriction* of R to Y is the set

$$\{(a, b) \in R : a \in Y \text{ and } b \in Y\}.$$

This is denoted by $R|Y$. This notion will be used in Chapter 6.

Theorem 3.3

If R is an order relation on a set X and $Y \subseteq X$ then $R|Y$ is an order relation on Y.

Proof

An easy exercise.

▶ As with any kind of mathematical structure, the idea of a structure-preserving function is important.

Definitions

Let X be ordered by a relation R and let Y be ordered by a relation S. A function $f : X \to Y$ is *order-preserving* if, for every $a, b \in X$, $(a, b) \in R$ if and only if $(f(a), f(b)) \in S$.

An *order isomorphism* is an order-preserving bijection. Two ordered sets are *isomorphic* if there is an order isomorphism between them.

Examples 3.4

(a) Let X denote the set of odd integers and let Y denote the set of even integers.

   The sets X and Y are ordered by magnitude: $R = \{(a, b) \in X \times X : a \leq b\}$, $S = \{(c, d) \in Y \times Y : c \leq d\}$. Also X and Y are isomorphic as ordered sets via the function f such that $f(x) = x + 1$ (there are other isomorphisms also).

(b) Take $\mathbb{R}$, ordered by magnitude, and the open interval $(-\pi/2, \pi/2)$, also ordered by magnitude. Then these are isomorphic as ordered sets, via the tangent function.

(c) Let $\mathbb{Z}^-$ be the set of negative integers and let $R = \{(x, y) \in \mathbb{Z}^- \times \mathbb{Z}^- : y \leq x\}$. Note that R is the converse of the standard order by magnitude. Then $\mathbb{Z}^-$ with ordering R is isomorphic to $\mathbb{Z}^+$, ordered by magnitude. The function which takes $-x$ to x is an isomorphism, since we have $-y \leq -x$ if and only if $x \leq y$.

▶ Generally speaking, an order isomorphism preserves all *order* properties, and it is often easy thereby to judge intuitively whether two ordered sets are isomorphic. Certainly, if the answer is negative, it may be seen to be so by observing a single characteristic which is not preserved. Of course, equinumerosity is a prerequisite. Also, existence of least, greatest, maximal or minimal elements may be a guide. Further, whether the order is partial or total will be relevant. Let us now turn our attention to these new ideas.

The notions of least and greatest elements are self evident on the analogy of a general order relation with a familiar order by magnitude (for example on $\mathbb{Z}$). If R is an order relation on a set X and aRb, we shall think of a as being 'smaller' than b, although the definition of the particular R may have nothing whatever to do with 'size'. Thus, a is the *least element* of X if aRx for every $x \in X$ and similarly b is the *greatest element* of X if xRb for every $x \in X$.

The first thing to observe about least and greatest elements is that there may not be any, and there can be two reasons for this. The first is exemplified by the set $\mathbb{Z}$ ordered by magnitude, where it is the infinity of elements which ensures no greatest or least elements. The second is exemplified by the set of all *proper* subsets of the set $\{1, 2, 3\}$, ordered by inclusion. Here the set is finite, but there is no element greater than every other element. This example leads us to the more interesting and more significant ideas of minimal and maximal elements.

Intuitively, an element of an ordered set is minimal if there is no smaller element in the set, and maximal if there is no larger element. In more formal terms, the definition is as follows.

Definition

Let A be a set ordered by the relation R. An element a of A is *minimal* if xRa implies $x = a$, for every $x \in A$. An element b of A is *maximal* if bRx implies $b = x$, for every $x \in A$.

Theorem 3.5

Let A be ordered by the relation R. If a is the least element of A then a is minimal and there is no other minimal element. Likewise, if b is the greatest element of A then b is maximal and there is no other maximal element.

Proof

Let a be the least element of A and let $x \in A$ with xRa. Since a is least, we must have aRx, so by the anti-symmetry of R we deduce $a = x$. Hence, a is minimal. Now suppose that a' is a minimal element. Since a is least in A, we know that aRa'. But by the definition of a minimal element this implies that $a = a'$. Thus there can be no minimal elements except a.

The proof for greatest and maximal elements is exactly analogous and is left as an exercise.

► To see the distinctions between least and minimal and between greatest and maximal it is best to consider examples, and a geometrical intuition about ordered sets can be helpful, so we shall attempt to develop this.

Examples 3.6

(a) Let $A = \{2, 3, 4, 6, 8, 12\}$ and let xRy if and only if x divides y. We construct a diagram (Fig. 3.1) in which the elements of A are represented by points, and whenever xRy we ensure that x lies below y and is joined to it by an upward path. Notice, for example, that 2 and 3 are not related by R, nor are 6 and 8. But although there is no direct line joining 2 with 8, the line via 4 serves to indicate in the diagram that $2R8$, since any ordering is necessarily transitive.

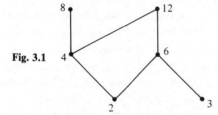

Fig. 3.1

This ordered set has no least element and no greatest element. It has two minimal elements (2 and 3) and two maximal elements (8 and 12).

(b) Take A to be $\{1, 2, 3, 4, 6, 8, 12, 24\}$ (adjoin 1 and 24), with ordering by 'divides' as above. Then 1 is the least element and 24 is the greatest element, and there are no other minimal or maximal elements (see Fig. 3.2).

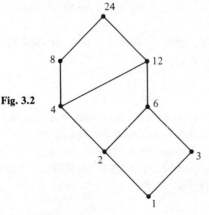

Fig. 3.2

(c) The set of all proper subsets of the set $\{1, 2, 3\}$, ordered by $\subseteq$. Here $\emptyset$ is the least element, and there are three maximal elements: $\{1, 2\}$, $\{2, 3\}$, and $\{3, 1\}$. See Fig. 3.3.

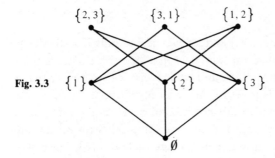

Fig. 3.3

(d) The set of all proper subsets of $\mathbb{N}$, ordered by $\subseteq$. This is an infinite set, so we cannot draw a diagram, but we can imagine how the diagram above would extend to this case. $\emptyset$ will be the least element, at the bottom of the diagram. There will be infinitely many singleton sets on the next level up, pairs on the next, triples on the next, etc. At the top of the diagram we shall have infinitely many maximal sets: $\mathbb{N}\backslash\{0\}$, $\mathbb{N}\backslash\{1\}$, $\mathbb{N}\backslash\{2\}$, etc.

(e) The set of all (non-empty) linearly independent sets of vectors in $\mathbb{R}^3$, ordered by $\subseteq$. Again this is infinite, but we can visualise what happens at the bottom and the top of the diagram. All the singleton sets $\{v\}$ are minimal, provided $v \neq 0$ (since $\{0\}$ is not linearly independent). All the three-element linearly independent sets $\{u, v, w\}$ are maximal, since we know that no linearly independent subset of $\mathbb{R}^3$ can have more than three elements, and every two-element set can be extended. These three-element sets are all bases for $\mathbb{R}^3$. (A basis is a maximal linearly independent subset.)

(f) $\mathbb{Z}$, ordered by magnitude. Here there are no least, greatest, minimal or maximal elements.

Theorem 3.7

An order isomorphism preserves least, greatest, minimal and maximal elements. More precisely, let A be ordered by R, let B be ordered by S, and let $f : A \to B$ be an order isomorphism. Then $a \in A$ is least (respectively, greatest, minimal, maximal) if and only if $f(a)$ is least (respectively greatest, minimal, maximal) in B.

Proof

Let a be least in A. Let $y \in B$. Then $y = f(x)$ for some $x \in A$ since f is a bijection. Now aRx since a is least in A, so $f(a)Sf(x)$, i.e. $f(a)Sy$. Thus $f(a)$ is least in B. Now let $f(a)$ be least in B, and let $u \in A$. $f(a)Sf(u)$ since $f(a)$ is least, so aRu since f is an order isomorphism. Thus a is least in A, as required.

The other three proofs (for greatest, minimal and maximal) are similar and are left as exercises.

▶ Again let us see the application of this result by means of examples.

Examples 3.8

(a) $\mathbb{Z}$ and $\mathbb{N}$ (both ordered by magnitude) cannot be isomorphic, since $\mathbb{N}$ has a least element, and $\mathbb{Z}$ does not.

(b) There can be no order isomorphism between the sets $A = \{1, 2, 3, 4, 6, 8, 12\}$ and $B = \{2, 3, 4, 6, 8, 12, 24\}$, both ordered by 'divides', since A has a least element and B has not (also B has a greatest element and A has not).

(c) The set of all proper subsets of $\{1, 2, 3\}$, ordered by inclusion, and the set A ordered by 'divides' from (b) above cannot be isomorphic, as is apparent from their diagrams (see Fig. 3.4). Both have a least element and neither has a greatest. But one has *three* maximal elements while the other has only two.

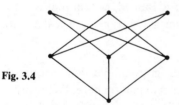 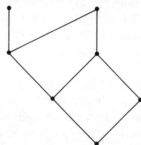

Fig. 3.4

(d) The sets $P(\mathbb{N})$ and $P(\mathbb{Z})$, both ordered by $\subseteq$. These are isomorphic. To see this we must construct an order isomorphism. Let $g : \mathbb{N} \to \mathbb{Z}$ be a bijection (see Corollary 2.11) and define $f : P(\mathbb{N}) \to P(\mathbb{Z})$ by $f(X) = \{g(x) : x \in X\}$. Then f is a bijection (see Exercise 5 on page 95). We must show that $X \subseteq Y$ if and only if $f(X) \subseteq f(Y)$, for all $X, Y \in P(\mathbb{N})$. Let $X \subseteq Y$ and let $a \in f(X)$. Then $a = g(x)$ for some $x \in X$. But then $x \in Y$ also, and so $a \in f(Y)$. We therefore have $f(X) \subseteq f(Y)$. Now suppose that $f(X) \subseteq f(Y)$ and let $x \in X$. Then $g(x) \in f(X)$, so $g(x) \in f(Y)$, i.e. $g(x) = g(y)$

for some $y \in Y$. But g is a bijection, so we must have $x = y$, and consequently $x \in Y$. Hence $X \subseteq Y$, as required.

Note that $P(\mathbb{N})$ and $P(\mathbb{Z})$ are both uncountable ordered sets. Each has a least element and a greatest element. Their diagrams have 'levels' just as in Example 3.6(d), consisting of singletons, pairs, triples, etc., and an order isomorphism will necessarily preserve these levels.

(e) The set $\mathbb{R}$, ordered by magnitude, is equinumerous with $P(\mathbb{N})$ but it is not isomorphic with $P(\mathbb{N})$ ordered by $\subseteq$. This follows from the existence of a least element in $P(\mathbb{N})$ and the non-existence of a least element in $\mathbb{R}$. However, there is another aspect of this example which is very significant. That is that for any two elements x and y of $\mathbb{R}$ we have either $x \leqslant y$ or $y \leqslant x$, whereas for any two elements X and Y of $P(\mathbb{N})$ we may have neither $X \subseteq Y$ nor $Y \subseteq X$.

This leads us to our next new idea.

Definition

An order relation R on a set X is a *total order* on X if, for every pair $x, y \in X$, we have either xRy or yRx.

It helps to consider what this means in terms of our diagrams. For any pair of points, one must lie below the other and be joined to it by a line (possibly with other points in between). And in turn this means that the points must all lie on a single line. Indeed, a term that is often used instead of 'total order' is '*linear* order' and this geometrical picture is the reason for it.

▶ In many books the term 'partial order' is used. We shall not use it, but it is sufficiently widely used to require explanation here. Every order relation (as we have defined the term) is a 'partial order' and every ordered set is a 'partially ordered set'. The word 'partial' is inserted to emphasise the possibility that there may exist pairs of elements which are not related by the order relation (i.e. the possibility which is excluded in the definition of a 'total' order). The need for this emphasis arose historically in the development of these matters – the notion of total order came first and was subsequently generalised to that of partial order. Nowadays the emphasis is unnecessary, and we can simply use the word 'order' to include both partial and total orders.

Examples 3.9

(a) $\mathbb{N}$, $\mathbb{Z}$ and $\mathbb{R}$ (ordered by magnitude) are examples of totally ordered sets.

(b) The set {1, 2, 3, 4, 6, 8, 12, 24}, ordered by 'divides' is not totally ordered, since (for example) 6 and 8 are not related. However, the set {1, 2, 4, 12, 24} is totally ordered by the relation 'divides'.

(c) Let R be the relation on $\mathbb{Z} \times \mathbb{Z}$ defined by: $(a, b)R(x, y)$ if $a \leq x$ and $b \leq y$. Then R is not a total order, since (for example) $(1, 2)$ and $(2, 1)$ are not related.

(d) $\mathbb{Z} \times \mathbb{Z}$ can be totally ordered. One total ordering is given by the relation S given as follows. $(a, b)S(x, y)$ if $a < x$ or if $a = x$ and $b \leq y$. A moment's reflection is all that is required to verify that any two pairs are related in S one way or the other.

► Example (b) above provides a good pictorial representation for the next definition. An ordered set in general will have many totally ordered subsets. For example {1, 2, 4, 12, 24} in the set {1, 2, 3, 4, 6, 8, 12, 24} (see Fig. 3.5). The subset inherits the ordering, and we have chosen it so that it is totally ordered. Another such subset is {3, 6, 12, 24}.

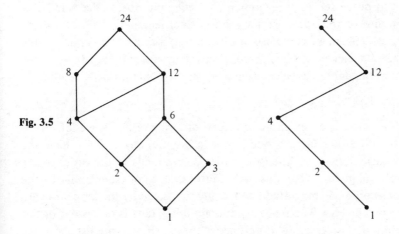

Fig. 3.5

Definition
    A *chain* in an ordered set is a subset which is totally ordered by the inherited order relation.

Of course, a chain need not be finite as in the above example. In $P(\mathbb{N})$, ordered by $\subseteq$, the collection of all sets of the form $\{1, 2, 4, \ldots, 2^n\}$, for different values of $n \in \mathbb{N}$, is an infinite totally ordered subset.

As is certainly suggested by our diagrams, it is the case that a totally ordered set cannot be isomorphic to an ordered set in which there are incomparable pairs.

Theorem 3.10

If A, totally ordered by the relation R, and B, ordered by the relation S, are isomorphic, then S is a total order on B.

Proof

Let $f: A \to B$ be an order isomorphism, and let $u, v \in B$. Then there exist $x, y \in A$ such that $u = f(x)$ and $v = f(y)$. Now R is a total order, so either xRy or yRx. Since f is an isomorphism, then, either $f(x)Sf(y)$ or $f(y)Sf(x)$, i.e. uSv or vSu. Hence, S is a total order on B.

► Amongst the totally ordered sets there is a class which will later receive particular attention at some length. These are the well-ordered sets, and they are important because of the link between well-ordering and counting. A counting process has a beginning and proceeds in discrete steps. As we saw in Chapter 2, we can generalise the idea of counting to some infinite sets, 'counted' by an infinite counting process. Later in the book we shall generalise this again and consider 'transfinite' counting processes. For the moment, one example must suffice to indicate the nature of this idea. Besides the normal listing 1, 2, 3, ... of the positive integers, we could provide a different listing by first enumerating all the odd elements of $\mathbb{Z}^+$, and, presuming that infinite list as given, enumerating all the even positive integers. We obtain the double list

$$1, 3, 5, 7, \ldots, 2, 4, 6, \ldots.$$

This of course is an example of a total ordering of $\mathbb{Z}^+$. If we let xRy if and only if either (a) x is odd and y is even, or (b) x and y have the same parity and $x \leqslant y$, then this order corresponds to the order in the above double list. That this ordering of $\mathbb{Z}^+$ is not isomorphic to the standard order by magnitude is strongly suggested by the 'appearances' of the two listings. The reader should fill in the details of a proof of this for himself. The essential difference is that in the double list there are *two* elements (namely, 1 and 2) which have no immediate predecessor.

The properties of the order relations corresponding to such generalised counting processes derive from the discrete counting steps and the 'directional' nature of the enumeration (the incomplete sequences are denoted by three dots leading to the *right*, i.e. *upwards* in the ordering). These properties are embodied in the next definition.

Definition

Let X be a set and let R be a total order relation on X. Then R *well-orders* X if every non-empty subset of X contains a least element.

Remarks 3.11

Let X be well-ordered by the relation R. Then

(a) X contains a least element (counted first).

(b) Each element of X other than the greatest element (if there is one) has an immediate successor. For if we choose $x \in X$ and let $A = \{y \in X : xRy \ \& \ x \neq y\}$, then A is non-empty (provided that x is not the greatest element of X) and so contains a least element.

(c) X contains no infinite descending chain, i.e. X contains no subset which may be represented in a list thus: $\ldots, x_3, x_2, x_1$, with $x_{i+1}Rx_i$ for each i. Clearly, any such subset would have no least element.

Examples 3.12

(a) $\mathbb{N}$, ordered by magnitude, is a well-ordered set. We shall not justify this – it is a consequence of Peano's axioms.

(b) Let A be any infinite countable set. Then there is a bijection $f : \mathbb{N} \to A$, say. This bijection gives rise to a well-ordering of A by: aRb if $a = f(m)$, $b = f(n)$ and $m \leq n$. Then R is then the image (under f) of the ordering of $\mathbb{N}$ by magnitude and f is an order isomorphism.

(c) $\mathbb{Z}$ is not well-ordered, under the standard order by magnitude.

(d) $\mathbb{Q}$, the set of rational numbers, ordered by magnitude, is not well-ordered, since (for example) it does not contain a least element. Consider then the set of all non-negative rational numbers, ordered by magnitude. This is not well-ordered either, since (for example) $\{q \in \mathbb{Q} : q > 1\}$ contains no least element. Notice, however, that $\mathbb{Q}$ can be given a well-ordering as in (b) above, since it is countable.

(e) Likewise, neither $\mathbb{R}$ nor any interval in $\mathbb{R}$ is well-ordered when ordered by magnitude.

(f) $\mathbb{N} \times \mathbb{N}$ is well-ordered by R, where $(a, b)R(x, y)$ if $a < x$ or if $a = x$ and $b \leq y$. Let us suppose that we have already verified that R is a total order, and now must show that it well-orders $\mathbb{N} \times \mathbb{N}$. Let X be a non-empty subset of $\mathbb{N} \times \mathbb{N}$. Denote by X_1 the set $\{a \in \mathbb{N} : \text{there is } b \in \mathbb{N} \text{ with } (a, b) \in X\}$. Since $\mathbb{N}$ is well-ordered by magnitude, X_1 contains a least element, say a_0. Now consider the set of elements of X of the form (a_0, b). The set of all $b \in \mathbb{N}$ such that $(a_0, b) \in X$ is a non-empty subset of $\mathbb{N}$, so contains a least element, say b_0. Then (a_0, b_0) is the least element of X, for if $(x, y) \in X$ and $(x, y)R(a_0, b_0)$ then either $x < a_0$, which is impossible by choice of a_0, or $x = a_0$ and $y \leqslant b_0$, which is possible only if $y = b_0$, by choice of b_0. Thus X contains a least element, as required.

Theorem 3.13

Let X, ordered by R, and Y, ordered by S, be ordered sets and let $f : X \to Y$ be an order isomorphism. If R is a well-ordering of X then S is a well-ordering of Y.

Proof

An exercise for the reader.

▶ It should be noted that the examples given above of well-ordered sets are all countable, so the question arises: is there a convenient example of an uncountable well-ordered set? The answer is: not for us at the moment, anyway. In particular, there is no convenient way in which to well-order the set $\mathbb{R}$ of real numbers. The proposition that $\mathbb{R}$ *can* be well-ordered was the subject of bitter disagreement amongst mathematicians in the early part of this century. The proof of this proposition (which was given first by Zermelo in 1904) does not take the form of an explicit description of a relation on $\mathbb{R}$ and verification that it well-orders $\mathbb{R}$. It is merely a demonstration that some such well-ordering must exist. The result and, consequently, the method of proof, were strongly disputed by some mathematicians who could not conceive of the continuum being well-ordered. Even today there is no way of specifying a well-ordering of $\mathbb{R}$. The reason for this is that the proof of existence (see Theorem 5.17) makes *essential* use of the axiom of choice, and this axiom is non-constructive, that is to say it is a mere assertion of existence. If it were possible to describe or construct explicitly a well-ordering of $\mathbb{R}$, this would amount to a demonstration that the use of the non-constructive axiom of choice was not essential. We shall return more than once to this point, first in Chapter 5 in discussion of the axiom of choice itself, and second in Chapter 6 in the context of ordinal and cardinal numbers. In Section 6.1 we shall give a description of a well-ordering of an uncountable set without the assumption of the axiom of choice, but in that case it will not be $\mathbb{R}$ that is well-ordered and, as we shall see, that example will be of no use in dealing with the question of the well-ordering of $\mathbb{R}$.

Exercises

1. Which of the following sets are ordered by the given relations?
   (i) $\mathbb{Z}^+$, where for $a, b \in \mathbb{Z}^+$ we have aRb if and only if a divides b.

(ii) $\mathbb{Z} \times \mathbb{Z}$, where for a, b, x, $y \in \mathbb{Z}$ we have $(a, b)R(x, y)$ if and only if $a \leqslant x$ and $b \leqslant y$.

(iii) $\mathbb{Z} \times \mathbb{Z}$, where for a, b, x, $y \in \mathbb{Z}$ we have $(a, b)S(x, y)$ if and only if either (a) $a < x$ or (b) $a = x$ and $b \leqslant y$.

(iv) $\mathbb{C}$, where for z_1, $z_2 \in \mathbb{C}$, $z_1 R z_2$ if and only if the real part of z_1 is less than or equal to the real part of z_2.

(v) $\mathbb{C}$, where for z_1, $z_2 \in \mathbb{C}$, $z_1 S z_2$ if and only if $|z_1| \leqslant |z_2|$.

(vi) The set of all $m \times n$ matrices, where for any such matrices A and B, ARB if and only if $a_{ij} \leqslant b_{ij}$ for $1 \leqslant i \leqslant m$ and $1 \leqslant j \leqslant n$.

(vii) The set of all real square matrices, where for any such matrices A and B, ASB if and only if $\det A \leqslant \det B$.

2. Draw diagrams to represent the following ordered sets.
    (i) The set of all subsets of the set $\{1, 2, 3, 4\}$, ordered by $\leqslant$.
    (ii) The set of natural numbers from 1 to 25 inclusive, ordered by divisibility.
    (iii) The sets in Exercise 1 parts (ii) and (iii).
    (iv) The set of real numbers of the form $1 + [n/(n+1)](n \in \mathbb{Z}^+)$, ordered by magnitude.
    (v) The set of real numbers of the form $m + [n/(n+1)](m, n \in \mathbb{Z}^+)$, ordered by magnitude.

3. Let X be a set and let $f : X \to \mathbb{R}$ be a function. A relation R on X may be specified by: xRy if and only if $f(x) \leqslant f(y)$, $(x, y \in X)$. Prove that R is an order relation on X if and only if f is an injection.

4. Let X be a set with two elements. How many essentially distinct (i.e. non-isomorphic) orderings of X are there? Repeat for a set with three elements and a set with four elements. (Use diagrams.)

5. Let A and B be sets and suppose that $f : A \to B$ is a bijection. This f induces a bijection $F : P(A) \to P(B)$ by $F(X) = \{f(x) : x \in X\}$ $(X \in P(A))$. Show that F is an order isomorphism from $P(A)$ to $P(B)$, where the order in each case is $\subseteq$.

6. Under what circumstances, if any, can the set of all *proper* subsets of a set Y have a greatest element when ordered by inclusion?

7. Prove or disprove the following: if an ordered set A contains precisely one minimal element then that element is the least element of A.

8. Find all maximal and minimal elements in the ordered sets listed in Exercises 1 and 2 above.

9. Amongst the following ordered sets, which pairs are isomorphic?
    (i) $\mathbb{Z}$, ordered by magnitude.
    (ii) The set of all infinite sequences of 0's and 1's, where $(u_n)R(v_n)$ if and only if either $u_n = v_n$ for all n or there is a number k such that $u_k < v_k$ and $u_r = v_r$ for $r < k$.
    (iii) $\mathbb{R} \times \mathbb{R}^+$, where $(a, b)R(x, y)$ if and only if $|x - a| \leqslant y - b$.
    (iv) $\mathbb{R}$, ordered by magnitude.
    (v) The set of all real numbers of the form $1 \pm [n/(n+1)](n \in \mathbb{Z}^+)$.
    (vi) The set of all open discs in the plane with centre lying on the x-axis, ordered by inclusion. (An open disc is the interior of a circle.)

10. Let X be a set and let R be a relation on X which is reflexive and transitive. For x, $y \in X$ let us say that $x \simeq y$ if xRy and yRx. Show that $\simeq$ is an equivalence relation on X. Denote the equivalence classes by

a bar (for example, $\bar{x}$). Define a relation $\bar{R}$ on the set of equivalence classes by $\bar{x}\bar{R}\bar{y}$ if and only if xRy ($x, y \in X$). Show that this is well-defined (i.e. if $\bar{x} = \bar{a}$ and $\bar{y} = \bar{b}$ then xRy if and only if aRb). Finally, show that $\bar{R}$ is an order relation on the set of equivalence classes. (Such a relation R is called a pre-order or quasi-order relation. Which of the relations listed in Exercise 1 above are pre-order relations but not order relations?)

11. Let X be a set and let $\mathcal{O}$ be the set of all relations on X which are order relations. $\mathcal{O}$ is itself ordered by $\subseteq$. Show that R is a maximal element of $\mathcal{O}$ if and only if R is a total order. (The question of whether $\mathcal{O}$ always contains a maximal element will be considered later. See Exercise 5 on page 183.)

12. Let X be a set and let $\mathcal{E}$ be the set of all relations on X which are equivalence relations. $\mathcal{E}$ is ordered by $\subseteq$. Investigate the maximal and minimal elements of $\mathcal{E}$ (if any). Is $\mathcal{E}$ necessarily totally ordered by $\subseteq$?

13. Find two ordered sets each of which is order isomorphic to a subset of the other but which are not themselves order isomorphic.

14. Prove that the double list ordering of $\mathbb{Z}^+$ described on page 92 is not isomorphic to the ordering of $\mathbb{Z}^+$ by magnitude.

15. Which of the following sets are well-ordered by the given relations?
    (i) The sets in Exercise 2 parts (iv) and (v) above.
    (ii) The set given in Exercise 9 part (ii) above.
    (iii) The collection of all subsets of $\mathbb{N}$ of the form $\{1, 2, 4, \ldots, 2^n\}$ (for $n \in \mathbb{N}$), ordered by inclusion.
    (iv) The collection of all subsets X of $\mathbb{N}$ such that all elements of X are powers of 2, ordered by inclusion.

16. Let X, ordered by R, and Y, ordered by S, be ordered sets, and let $f : X \to Y$ be an order isomorphism. Prove that if R well-orders X then S well-orders Y.

17. Let X and Y be well-ordered sets. Describe one way of constructing a well-ordering of $X \times Y$. Extend this to Cartesian products of any finite number of well-ordered sets.

18. Let X be well-ordered by the relation R and suppose that X contains a greatest element. Is the converse relation R^{-1} necessarily a well-ordering of X? If not, find a condition on X and R which will ensure that R^{-1} is a well-ordering.

19. Let X be well-ordered by the relation R, and let $f : X \to X$ be an order-preserving injection. Show that $xRf(x)$ for every $x \in X$. Deduce that there cannot exist $x_0 \in X$ such that X is order isomorphic with $\{x \in X : xRx_0 \text{ and } x \neq x_0\}$, ordered by the restriction of R (i.e. X cannot be order isomorphic with a proper initial segment of itself).

20. Let X be a set totally ordered by the relation R. A subset Y of X is said to be *cofinal* in X if for each $x \in X$ there exists $y \in Y$ with xRy. Show that X has a finite cofinal subset if and only if X has a greatest element. Find well-ordered cofinal subsets (minimal if possible) in each of the following sets, totally ordered by magnitude.
    (i) $\mathbb{N}$.   (ii) $\mathbb{Z}$.   (iii) $\{x \in \mathbb{N} : x < 100\}$.

(iv) $\{x \in \mathbb{Z} : x < 100\}$.   (v) $\mathbb{R}$.
(vi) $\{x \in \mathbb{R} : x \le 0\}$.   (vii) $\{x \in \mathbb{R} : x < 0\}$.
(See Exercise 12 on page 184.)

3.2 Lattices and Boolean algebras

Ordered sets will play an important part in the remainder of this book. The most significant applications are in Chapter 5 with regard to Zorn's lemma, and in Chapter 6 with regard to ordinal numbers and transfinite induction. The ideas of the previous section will be essential for these chapters. However, for the moment we shall digress in order to describe another special kind of ordered set which leads to another area of mathematics. This is lattice theory, and we shall give the sketchiest of introductions to the basic notions.

Definition

A *lattice* is an ordered set in which every pair of elements has a least upper bound and a greatest lower bound. More precisely, if A, ordered by R, is to be a lattice then for every $x, y \in A$ the set $U = \{u \in A : xRu \text{ and } yRu\}$ must have a least element and the set $V = \{v \in A : vRx \text{ and } vRy\}$ must have a greatest element.

The greatest lower bound of x and y is denoted by $x \wedge y$ and the least upper bound by $x \vee y$.

Examples 3.14

(a) For any set X, $P(X)$, ordered by $\subseteq$, is a lattice. Here $\wedge$ is intersection and $\vee$ is union.

(b) $\mathbb{Z}$, ordered by magnitude, is a lattice. Here $x \wedge y$ is $\min(x, y)$ and $x \vee y$ is $\max(x, y)$. Similarly, any totally ordered set is a lattice.

(c) $\mathbb{R} \times \mathbb{R}$, ordered by the relation R given by $(a, b)R(x, y)$ if and only if $a \le x$ and $b \le y$, is a lattice. Here

$$(a, b) \wedge (x, y) = (\min(a, x), \min(b, y)),$$

$$(a, b) \vee (x, y) = (\max(a, x), \max(b, y)).$$

(d) On first thinking about it, it may be hard to conceive of an ordered set which is not a lattice. However, it is quite easy to construct an example. Let S be the relation on $\mathbb{Z} \times \mathbb{Z}$ defined by $(a, b)S(x, y)$ if and only if $a = x$ and $b \le y$. Verification that S orders $\mathbb{Z} \times \mathbb{Z}$ is left as an exercise. If $a \ne x$ then for any $b, y \in \mathbb{Z}$,

(a, b) and (x, y) have no upper bound and no lower bound in this ordering. Thus $\mathbb{Z} \times \mathbb{Z}$ is not a lattice when ordered by S.

(e) Another example which is not a lattice is the set of all discs in the plane, ordered by $\subseteq$. In this case upper and lower bounds exist but there are, in general, no least and greatest (respectively) amongst them.

► A totally ordered set is the 'thinnest' possible lattice, and as such is rather trivial. The principal example which motivated the development of lattice theory is a collection of sets closed under union and intersection, and we shall call this a *lattice of sets*. (It is easy to verify that such a collection, ordered by $\subseteq$, is necessarily a lattice.) This is not the most general amongst lattices, however, since a lattice of sets always has a particular property. It is necessarily distributive.

Definition

A lattice is *distributive* if, for all elements a, b, c of the lattice,

$$a \wedge (b \vee c) = (a \wedge b) \vee (a \wedge c),$$

and

$$a \vee (b \wedge c) = (a \vee b) \wedge (a \vee c).$$

It should be noted that the operations $\wedge$ and $\vee$ in any lattice have certain basic algebraic properties, for example they are both commutative and associative. The reader should convince himself of these. To consolidate these ideas it is a worthwhile exercise to demonstrate that the two distributive laws above are in fact equivalent; each implies the other.

In the case of a lattice of sets, $\wedge$ is $\cap$ and $\vee$ is $\cup$ and the two distributive laws above are certainly satisfied. These laws do not hold in every lattice, as the following example shows.

Example 3.15

Let $A = \{a, b, c, d, e\}$ and let R be the order relation represented in Fig. 3.6. A can be easily seen to be a lattice with this ordering. Now

$$c \vee (b \wedge d) = c \vee a = c,$$

and

$$(c \vee b) \wedge (c \vee d) = e \wedge e = e.$$

A is therefore a non-distributive lattice.

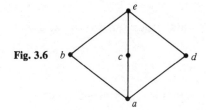

Fig. 3.6

▶ Distributive lattices are attractive to mathematicians because there is a nice representation theorem (Theorem 3.18) which says that every distributive lattice is order isomorphic to a lattice of sets. So the structure of any distributive lattice is 'contained' in some lattice of sets. We shall develop sufficient of the theory to see how a proof of this result works.

Definition

Let A, ordered by R, be a distributive lattice. A non-empty subset I of A is an *ideal* of A if (for $a, b \in A$)

(i) $a \in I$ and $b \in I$ together imply $a \vee b \in I$,

and

(ii) if $b \in I$ and aRb, then $a \in I$.

Note

There is a dual notion, which we define for the sake of completeness, though we shall not in fact have occasion to use it. A non-empty subset F of A is a *filter* of A if

(i) $a \in F$ and $b \in F$ together imply $a \wedge b \in F$,

and

(ii) if $b \in F$ and bRa, then $a \in F$.

All of our results about ideals can be translated into results about filters also.

▶ An ideal is closed under the least upper bound operation and closed under movement downwards in the ordering. Notice that if the lattice contains a least element a_0 then a_0 must be a member of every ideal.

It is not hard to show from the definition that an ideal is also closed under the greatest lower bound operation $\wedge$. Let a, b belong to an ideal I in a lattice A, ordered by R. $(a \wedge b)Rb$ always, so by (ii) in the definition of ideal, $a \wedge b \in I$. An immediate consequence of this is that an ideal in a lattice is itself a lattice under the inherited ordering. Similarly, a filter must be a lattice.

Examples 3.16

(a) For any distributive lattice A with order relation R, and any element $a \in A$, the set

$$\{x \in A : xRa\}$$

is an ideal of A, called the principal ideal generated by a.

(b) $P(X)$, ordered by $\subseteq$, for any given set X, is a distributive lattice. Take any element $x \in X$. Then the collection of subsets of X which do not contain x is an ideal in $P(X)$. For condition (i), we know that $x \notin U$ and $x \notin V$ together imply $x \notin U \cup V$. For condition (ii), we know that if $x \notin V$ and $U \subseteq V$ then $x \notin U$.

(c) $\mathbb{Z} \times \mathbb{Z}$, ordered by the relation R given by $(a, b)R(x, y)$ if and only if $a \leq x$ and $b \leq y$, is a distributive lattice (verify as an exercise). The set

$$\{(x, y) \in \mathbb{Z} \times \mathbb{Z} : x \leq 0 \text{ and } y \leq 0\}$$

is an ideal. For condition (i) of the definition, if $a \leq 0$, $b \leq 0$, $x \leq 0$, $y \leq 0$, then certainly $\max(a, x) \leq 0$ and $\max(b, y) \leq 0$. For condition (ii), if $x \leq 0$, $y \leq 0$ and $(a, b)R(x, y)$ then we must have $a \leq x \leq 0$ and $b \leq y \leq 0$. Note that this example is a particular case of (a) above. It is the principal ideal generated by $(0, 0)$.

Definition

An ideal I in a lattice A is a *prime ideal* if $I \neq A$ and for $a, b \in A$, $a \wedge b \in I$ implies either $a \in I$ or $b \in I$.

▶ Notice the analogy between this definition and one of the properties of prime *numbers*: if p is prime then, for any integers m and n, if mn is a multiple of p then either m is a multiple of p or n is a multiple of p. This analogy is more than a formal one, as the following example shows.

Example 3.17

Let the relation R be defined on $\mathbb{Z}^+$ by: aRb if and only if b divides a. Then $\mathbb{Z}^+$, with this ordering, is a distributive lattice (requires verification) and $\wedge$ and $\vee$ are given by:

$$a \wedge b = \text{least common multiple of } a \text{ and } b,$$

and

$$a \vee b = \text{greatest common divisor of } a \text{ and } b.$$

For any $x \in \mathbb{Z}^+$, the set $x\mathbb{Z}^+$ of all multiples of x is an ideal. For condition (i), $a = kx$ and $b = lx$ $(k, l \in \mathbb{Z}^+)$ implies that x is a common divisor of a and b, and hence the g.c.d. of a and b is a multiple of x also i.e. $a \vee b \in x\mathbb{Z}^+$. For condition (ii) if $b = kx$ $(k \in \mathbb{Z}^+)$ and aRb then $a = l(kx)$ for some $l \in \mathbb{Z}^+$, so $a \in x\mathbb{Z}^+$ also. Now suppose that p is prime. We show that $p\mathbb{Z}^+$ is a prime ideal. Let $a, b \in \mathbb{Z}^+$ be such that $a \wedge b \in p\mathbb{Z}^+$, i.e. the l.c.m. of a and b is a multiple of p. Then it is an easy deduction of elementary number theory that either $a \in p\mathbb{Z}^+$ or $b \in p\mathbb{Z}^+$, so $p\mathbb{Z}^+$ is a prime ideal. Elementary number theory also yields the converse to this: if $x\mathbb{Z}^+$ is a prime ideal then x is a prime number, but we leave this for the reader to verify.

Theorem 3.18
Every distributive lattice is isomorphic to a lattice of sets.

Proof
We give the merest sketch of the proof. Let A be a distributive lattice, ordered by the relation R. With each $a \in A$ we associate the set $\mathscr{I}(a)$ of all prime ideals of A which do not contain a. It can be shown that, for distinct $a, b \in A$, we must have $\mathscr{I}(a) \neq \mathscr{I}(b)$. It can further be shown that aRb if and only if $\mathscr{I}(a) \subseteq \mathscr{I}(b)$. Hence, $\mathscr{I}$ is in fact an order isomorphism from A to the lattice of sets $\{\mathscr{I}(a) : a \in A\}$.

▶ It should certainly be remarked at this stage that the details of the above proof are not easy. The reader is referred to the book by Kuratowski & Mostowski. In particular, it should be noted in relation to the material of Chapter 5 that the axiom of choice is used in the proof.

The example of a lattice of sets leads to another kind of mathematical structure more restrictive than the distributive lattice, and that is the Boolean algebra. The example here to bear in mind is the lattice of *all* subsets of a given set X. This lattice of course is distributive but it has some further significant properties not shared by all lattices of sets. First, it contains a least element (the empty set) and a greatest element (the whole set X). Second, for each element $U \in \boldsymbol{P}(X)$, there is a complement $V \in \boldsymbol{P}(X)$, i.e. an element V such that $U \wedge V = \emptyset$ and $U \vee V = X$.

Definition
A *Boolean algebra* is a distributive lattice A with a least element (say, 0) and a greatest element (say, 1) such that for each $x \in A$ there

is $x' \in A$ satisfying

$$x \wedge x' = 0 \quad \text{and} \quad x \vee x' = 1.$$

This x' is called the *complement* of x. It can be shown that complements are unique if they exist (see Exercise 10 on page 107).

Examples 3.19

(a) $P(X)$, ordered by $\subseteq$, is a Boolean algebra, for any set X, as noted above.

(b) The set $\{a, b, c, d\}$ with the ordering given by the diagram in Fig. 3.7 is a Boolean algebra. (Verification is left as an exercise, but the existence of least and greatest elements and of complements is clear.)

Fig. 3.7

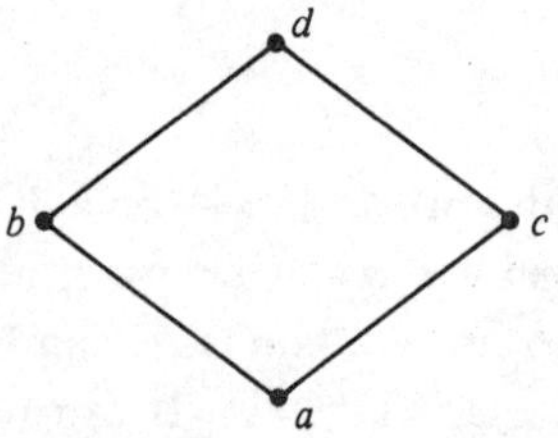

(c) A subset of a topological space is said to be clopen if it is both open and closed. The collection of all clopen subsets of a topological space T is a Boolean algebra, ordered by $\subseteq$.

(d) The collection of all subsets A of $\mathbb{Z}$, such that either A or $\mathbb{Z} \setminus A$ is finite, is a Boolean algebra, ordered by $\subseteq$. (Note that this is a countable Boolean algebra.)

(e) This example will be meaningful only to those readers who have studied some logic. In the calculus of propositions, we say that two propositions p and q are equivalent if both $p \Rightarrow q$ and $q \Rightarrow p$ are tautologies. The set of all equivalence classes of propositions is a Boolean algebra, where the ordering is given as follows (where $\bar{p}$ denotes the equivalence class of p):

$\bar{p} R \bar{q}$ if and only if $p \Rightarrow q$ is a tautology.

That R is well-defined is an elementary theorem of mathematical logic which we shall not discuss. We wish merely to present the example to illustrate that Boolean algebras occur in various branches of mathematics. Indeed, Boolean algebras occur frequently in mathematical logic.

▶ It is tempting to conjecture that every Boolean algebra might be isomorphic to a power set Boolean algebra, i.e. one of those in (a) above. That this cannot be so is demonstrated by (d) above and the knowledge (see Exercise 10 on page 63) that $P(X)$ cannot be infinite and countable, for any set X. However, there is a result along these lines which is a straightforward consequence of Theorem 3.18.

Theorem 3.20 (Stone representation theorem)
Every Boolean algebra is isomorphic to a Boolean algebra consisting of a collection of subsets of a set X, in which the order is $\subseteq$.

▶ Let us now examine some properties of Boolean algebras in general.

Remarks 3.21

(a) For any Boolean algebra A, in which the order relation is R, the converse relation R^{-1} also orders A (Theorem 3.2) and A, ordered by R^{-1}, is a Boolean algebra. The function which takes each element of A to its complement is an order isomorphism between these two Boolean algebras. An ideal in one is a filter in the other.

(b) An ideal in a Boolean algebra A cannot contain both an element a and its complement a' unless it consists of all of A. For suppose that I is an ideal of A and $a, a' \in I$. Then $a \vee a' \in I$, i.e. $1 \in I$. Now for every $x \in A$, we have $xR1$, so $x \in I$ since I is an ideal. Consequently, $I = A$.

(c) The intersection of any collection of ideals in a Boolean algebra A is an ideal in A. Verification is an easy exercise.

(d) An ideal I in a Boolean algebra A is *maximal* if $I \neq A$ and there is no ideal other than A itself which strictly contains I. We show that an ideal I in A is maximal if and only if, for each $a \in A$, either a or its complement belongs to I. First, suppose that for each $a \in I$, either $a \in I$ or $a' \in I$, and let J be an ideal containing I with $J \neq I$. Choose $b \in J \backslash I$. Since $b \notin I$, we must have $b' \in I$, so $b' \in J$ since $I \subseteq J$. But then both $b \in J$ and $b' \in J$, so $J = A$, by (b) above, and I must be maximal. Conversely, let I be a maximal ideal and suppose that $a \notin I$. Let K be the set of all elements of A of the form $b \vee x$, with bRa and $x \in I$ (where A is ordered by R). K is an ideal in A (verification left as an exercise), and $I \subseteq K$. Hence, $K = A$, and so for some b with

bRa and some $x \in I$ we have $b \vee x = 1$. Consequently, $a \vee x = (a \vee b) \vee x = a \vee (b \vee x) = a \vee 1 = 1$ and so

$$a' = a' \wedge 1 = a' \wedge (a \vee x) = (a' \wedge a) \vee (a' \wedge x) = 0 \vee (a' \wedge x) = a' \wedge x.$$

Now $(a' \wedge x)Rx$, so $a'Rx$, and $x \in I$. It follows that $a' \in I$, since I is an ideal. The result is now proved.

(e) An ideal I in a Boolean algebra A is prime if and only if it is maximal. This is quite a useful exercise. It uses the fact, true in all Boolean algebras, that for any two elements a and b, $(a \wedge b)' = a' \vee b'$ (this is a generalisation of the elementary rule for the complement of the intersection of two sets). Notice, however, that an ideal in a lattice can be prime without being maximal – take, for example, the set $\mathbb{Z}$ ordered by magnitude, and the ideal $\{x \in \mathbb{Z} : x \leq 0\}$, which is prime, but this lattice contains no maximal ideals. Further, an ideal in a lattice can be maximal without being prime. See Exercise 13 at the end of this section.

(f) Results similar to the above hold for filters. The order isomorphism described in (a) above is the means of demonstrating a duality in which $\wedge$ and $\vee$ are interchanged and 'filter' and 'ideal' are interchanged. Because of this duality, some authors use the term 'dual ideal' rather than 'filter'. We shall not pursue this but it may be useful to note that the term *ultrafilter* is commonly used in the literature to mean a maximal filter.

▶ The most significant theorem about Boolean algebras is the prime ideal theorem. This result has many consequences outside the field of Boolean algebras, some of which will be dealt with at greater length in Chapter 5.

Theorem 3.22 (Boolean prime ideal theorem)
    Every Boolean algebra contains a prime (equivalently, maximal) ideal.

We shall not prove this now. A proof will be given in Chapter 5, since it uses one of the results of that chapter, namely Zorn's lemma. As we shall see, Zorn's lemma is equivalent to the axiom of choice, so Theorem 3.22 is a consequence of the axiom of choice.

▶ There is a theorem corresponding to Theorem 3.22 for lattices in general. But before we state it let us recall that in general there is a difference between prime ideals and maximal ideals and that in general a lattice may contain neither a prime ideal nor a maximal ideal.

Example 3.23

Let A be the set of all finite open intervals in $\mathbb{R}$, ordered by $\subseteq$. Then A is a lattice. We shall take $\emptyset$ to be a member of A. Then A contains neither a prime ideal nor a maximal ideal. Notice first that for $X, Y \in A$, we have $X \wedge Y = X \cap Y$ and $X \vee Y =$ smallest open interval containing both X and Y (which, if X and Y are disjoint, is *not* $X \cup Y$). Suppose that I is a prime ideal, and that $I \neq A$, so let $X \in A \backslash I$. Since $\emptyset \in I$, if $Y \in A$ and $Y \cap X = \emptyset$, we must have $Y \in I$, by the primeness of I. Choose Y_1 and Y_2 both disjoint from X, one to the left of X on the real line and one to the right. Then $Y_1 \vee Y_2 \in I$ and $X \leqslant Y_1 \vee Y_2$, so $X \in I$. Contradiction. Hence I cannot be prime, and A contains no prime ideals.

Now suppose that J is a maximal ideal, and $J \neq A$, so let $X \in A \backslash J$ with $X = (a, b)$. By an argument similar to the above, all elements of J must lie on the same side (right or left) of X on the real line, i.e. we have either

$$Y \in J \quad \text{implies} \quad Y \subseteq (-\infty, b),$$

or

$$Y \in J \quad \text{implies} \quad Y \subseteq (a, \infty).$$

Suppose the former, without loss of generality. Then the set of all finite intervals contained in $(-\infty, b+1)$ is a proper ideal of A which properly contains J, contradicting maximality.

Note that this lattice does contain ideals. As always, there are principal ideals, namely, all open subintervals of a given finite interval, and there are others, for example all finite open subintervals of a given infinite interval.

Theorem 3.24

Every lattice which contains a greatest element and at least one other element has a maximal ideal.

Proof

We omit the proof, which again uses the axiom of choice. (See Exercise 5 on page 183.)

▶ To close this chapter it is right to mention the other approach that can be made to the structure embodied in a Boolean algebra. This will not have any significance for us, but should clarify the definitions given in many texts which appear to be rather different from ours.

Alternative definition

A Boolean algebra is a set A with two binary operations $\wedge$ and $\vee$ satisfying the following axioms.

(1) $\wedge$ and $\vee$ are both commutative and associative.

(2) There exist distinct elements 0 and 1 in A such that for each $a \in A$,

$$a \vee 0 = a \quad \text{and} \quad a \wedge 1 = a.$$

(3) The distributive laws hold.

(4) For each $a \in A$ there is an element $a' \in A$ such that

$$a \vee a' = 1 \quad \text{and} \quad a \wedge a' = 0.$$

It is apparent that there is no mention of an order relation in this definition. However, the order is there intrinsically and we can make the definition, for any $a, b \in A$;

$$aRb \text{ if and only if } a \wedge b = a.$$

It is a straightforward exercise to verify that R is an order relation and that for any $a, b \in A$, $a \wedge b$ and $a \vee b$ are in fact the greatest lower bound and least upper bound respectively.

The alternative definition has precedence historically. The original examples were the algebra of all subsets of a given set (with operations $\cap$ and $\cup$), and the algebra (customarily called the calculus) of propositions. Both examples have been mentioned already (Examples 3.19) but some further amplification of the latter is desirable. Two propositions p and q are equivalent if both $p \Rightarrow q$ and $q \Rightarrow p$ are tautologies. Equivalence classes are denoted by a bar, as in $\bar{p}$. The operations $\wedge$ and $\vee$ on the set of equivalence classes are given by:

$$\bar{p} \wedge \bar{q} = \overline{p \text{ and } q}$$

$$\bar{p} \vee \bar{q} = \overline{p \text{ or } q \text{ or both}}.$$

It can be verified that these are well-defined and that they satisfy the axioms for a Boolean algebra. The element 1 is the equivalence class of all tautologies, and 0 is the equivalence class of all contradictions.

George Boole (1815–64) was amongst the first to recognise the analogy between the algebra of sets and the calculus of propositions which is given substance in the definition of a Boolean algebra.

Exercises

1. Show that the following ordered sets are lattices.
   (i) Any totally ordered set.
   (ii) $\mathbb{Z}^+$, ordered by divisibility.
   (iii) The set of all subgroups of a given group, ordered by $\subseteq$.
   (iv) The set of all ideals in a given ring, ordered by $\subseteq$.
   (v) $\mathbb{R} \times \mathbb{R}$, ordered by R, where $(a, b)R(x, y)$ if and only if $a \leqslant x$ and $b \leqslant y$.
   (vi) The sets given in Exercise 9 on page 95.

2. Let A, ordered by R, be a lattice. Show that A, ordered by R^{-1}, is also a lattice. If the first lattice is distributive, can we deduce that the second is also?

3. Prove that, in any lattice, each distributive law implies the other.

4. Give an example of an infinite lattice which is not distributive.

5. Prove that in any lattice A, with order relation R, the set $\{x \in A : xRa\}$ is an ideal, for each element $a \in A$.

6. Prove that the intersection of any collection of ideals in a lattice is an ideal.

7. Let A be a lattice, with order relation R, and let I be an ideal in A. Prove that I is a filter in the lattice obtained by ordering A by the relation R^{-1}.

8. For each of the lattices given in Exercise 1 parts (i), (ii), (v) and (vi), find examples of prime ideals and non-prime ideals, if possible.

9. An ideal I in a lattice A is maximal if $I \neq A$ and there is no ideal other than A itself which strictly contains I. Give an example of a lattice with an ideal which is maximal but not prime.

10. Prove that if A is a Boolean algebra (with order relation R) and $a \in A$ then there is precisely one element $x \in A$ such that $x \wedge a = 0$ and $x \vee a = 1$, namely a'. Prove also that the function from A to A which takes each element to its complement is an injection. Finally, show that this function is an order isomorphism between A ordered by R and A ordered by R^{-1}.

11. Let A be a Boolean algebra and let $a, b \in A$. Prove that $(a \wedge b)' = a' \vee b'$.

12. Draw diagrams representing all Boolean algebras with fewer than nine elements.

13. Prove that an ideal in a Boolean algebra is prime if and only if it is maximal. (See Remark 3.21(e).)

Further reading

Birkhoff & Maclane [3] A standard textbook on abstract algebra, containing a section on lattices and Boolean algebras.

Kuratowski & Mostowski [18] A wide-ranging and useful reference book, though its treatment of some topics is rather different from ours. Quite technical.

Rutherford [21] An exposition of the standard results of lattice theory.

Stoll [24] A very readable book on logic and set theory, which includes a substantial section on Boolean algebras.

4

SET THEORY

Summary

After a discussion of what sets are useful for, a list is given of set operations and constructions which are in normal use by mathematicians. Then there is a complete list of the Zermelo–Fraenkel axioms, followed by discussion of the meaning, application and significance of each axiom individually, including reference to historical development. Normal mathematics can be developed within formal set theory, and the basis of this process is described. A system of abstract natural numbers is defined within ZF set theory and demonstrated to satisfy Peano's axioms. As an alternative to ZF, the von Neumann–Bernays system VNB of set/class theory is described and its usefulness and its relationships with ZF are discussed. Finally, some of the logical and philosophical aspects of formal set theory are described, including consistency and independence results.

The reader is presumed to be familiar with the algebra of sets and with standard set constructions and notation. Some experience with abstract algebraic ideas is useful. Section 1.1 is referred to, but this chapter is essentially independent of Chapters 2 and 3. No knowledge of mathematical logic is assumed.

4.1 **What is a set?**

On the face of it, the notion of set is one of the simplest ideas there can be. It is this simplicity and freedom from restrictive particular properties which make the notion so suitable for use in abstract mathematics. Indeed, 'set' itself is an abstraction which means little in isolation. Taking this to the extreme, it may be argued that use of the term 'set' is nothing more than a way of speaking. Consider the following two statements:

(i) The set of all orthogonal real matrices is a subset of the set of all invertible real matrices.

(i) Every orthogonal real matrix is invertible.

These two statements mean the same. As another example, consider:

(iii) The set of all skew-symmetric invertible real matrices is empty.

(iv) No skew-symmetric real matrix is invertible.

These two statements mean the same. In these cases the use of the word 'set' and the notion 'empty' are inessential. The meaning can be expressed without them.

All of modern mathematics uses the terminology of set theory. For the most part, this use is inessential, in the same way as above, although translation out of the language of sets would frequently be very complicated. One field where this translation would of course not be possible is the study of set theory itself. That many mathematicians have spent many years investigating set theory for its own sake is substantial evidence that 'set' is indeed more than a way of speaking. The above discussion is intended to suggest, however, that the significance of sets is due to their *usefulness*, and that the claim that set theory is the essential basis for mathematics is an extravagant one.

The usefulness of sets is a modern phenomenon, and it has arisen along with the general abstraction of mathematics. This process started seriously in the nineteenth century and came through unconscious development to the explicit and unconstrained definitions of Cantor and the remarkably far sighted axiom system of Zermelo which ushered in the twentieth century.

In dealing with objects of a certain kind (say natural numbers) the mathematician comes across properties which objects of that kind may or may not have. For example, primeness is a property of some natural numbers. Thus the notion of 'all prime numbers' comes on the scene, and the mathematician considers as a single object the collection of all prime numbers. The familiar notation $\{x : x \text{ is a prime number}\}$ is the standard way of denoting the set determined by a property in this way. This correspondence between sets and properties has been commented on previously in relation to Boolean algebras. The algebra of sets is based on (or may be regarded as an expression of) the logic of propositions. Let us be absolutely clear about this, as it is fundamental. The reader is assumed to be familiar with the algebra of sets, but perhaps not with logic, so let us examine some examples.

Examples 4.1

(Each set is presumed to be a set of natural numbers.)

(a) $\{x : x^2 \text{ is even}\} \subseteq \{x : x \text{ is even}\}$.

The relationship of inclusion between sets is another way of expressing a logical relation between the propositions 'x^2 is even' and 'x is even', namely, the conditional: *if* x^2 is even *then* x is even.

(b) $\{x : x \text{ is prime}\} \cap \{x : x \text{ is even}\} = \{2\}$.

The properties here are primeness and evenness. Another way of expressing the above is:

if x is prime *and* x is even, *then* $x = 2$, and conversely.

The intersection of sets corresponds to the logical conjunction of propositions, and the equality of sets corresponds to the (logical) equivalence of propositions.

(c) $\{x : 9 \text{ divides } x\} \cup \{x : 12 \text{ divides } x\} \subseteq \{x : 3 \text{ divides } x\}$.

This is equivalent to:

if 9 divides x *or* 12 divides x, *then* 3 divides x.

(d) $\{x : x \text{ is prime}\} \subseteq \mathbb{N} \backslash \{x : x + 3 \text{ is prime}\}$.

This is equivalent to:

if x is prime *then* $x + 3$ is *not* prime.

Notice that the complement of a set corresponds to the negation of the proposition. (In this example the assertion is in fact false, of course.)

► This correspondence between the algebra of sets and logic was apparent to the nineteenth-century mathematicians who followed Boole. Up to a point (indeed, for most practical mathematical purposes) there are no difficulties about it. However, the study of sets in the abstract and attempts to list properties of abstract sets did lead to problems. The remaining contents of this book are largely devoted to a description of the sorts of problems that arose and the mathematics that was created in response to them.

It is in the nature of mathematics that precision is required about the meanings and properties of the words used. Thus it was perfectly natural for the uncomplicated nineteenth-century notion of set embodied in $\{x : x \text{ has property } P\}$ to be questioned. As soon as the question 'what is a set' is asked, we are into the area of abstract set theory and awkward questions can arise.

One such problem arises immediately from the kind of set construction referred to in the examples above. This is what is commonly known as *Russell's paradox*. Before discussing this we should perhaps make more explicit the set construction procedure. The correspondence between sets and properties became known as the *comprehension principle*, and may be stated as follows:

> Given any property, there is a set consisting of all objects which have that property.

This principle lay behind the introduction of the use of sets in the nineteenth century and clearly is the basis of the notation $\{x : x$ has property $P\}$. On intuitive grounds, it is hard to see what could go wrong with this. The comprehension principle expresses what is meant by a set in normal usage. In 1901, however, Russell made the following crucial observation.

Theorem 4.2
The comprehension principle leads to a contradiction.

Proof
Sets have elements. Indeed, sets may have other sets as elements. Hence, given any two sets x and y, it is reasonable to ask whether x is an element of y. There will always be a definite answer, yes or no. More particularly, given any set x, it is reasonable to ask whether x is an element of x. 'x belongs to x' may be true for some sets x and false for others.

Let us apply the comprehension principle to the property: 'x is not an element of x'. We obtain the set $\{x : x \notin x\}$. Denote this set by A. Now it must be the case that $A \in A$ *or* $A \notin A$. If $A \in A$ then A satisfies the requirement for belonging to the set A so $A \notin A$, and from this contradiction we conclude that the case $A \in A$ cannot obtain.

The other possibility must obtain, therefore, namely, $A \notin A$. But then A satisfies the requirement for belonging to A, and so $A \in A$. Again there is a contradiction, and this time we have no other possibilities to fall back on. A genuine contradiction is derivable from the comprehension principle as stated above.

▶ Another sort of problem is what to do about results such as the well-ordering theorem (see page 172). This states that given any set X

there is a relation which well-orders X. Cantor believed this to be true. His opponents believed it to be false, largely on the grounds that no way of well-ordering the set $\mathbb{R}$ of real numbers was apparent. It is a question about sets in the abstract. How can such questions be approached, and on what can judgments and deductions about such matters be based?

The answer to both these difficulties lies in the idea of a common starting point, as discussed in Chapter 1. Mathematicians need to agree as far as possible on the properties of sets, in order to make sensible use of the notion. And this is where the axioms for set theory come in. We do not *define* the term 'set' (just as we did not define the term 'number'). We agree on certain principles by means of which the *properties of sets* are characterised.

The formulation of a list of such principles was Zermelo's achievement in 1908 and, remarkably, his original list of axioms has been modified only slightly (though significantly) to yield the most widely accepted of today's formal set theories. Later in this chapter we shall examine two of these, the Zermelo–Fraenkel theory and the von Neumann–Bernays theory.

Before listing these formal axioms however, let us finish this section by a discussion of the sorts of properties that are going to have to be either expressed explicitly amongst the axioms or implied as consequences of them.

Remarks 4.3

(a) *Membership*

Sets have elements, and intuitively we can allow objects of any sorts to be collected together in a set. Indeed, sets can be elements of other sets.

Sets are equal only if they are identical. What this means is that they have the same elements. Of course, the same set may be represented in different ways, however. For example, $\{0, 1\}$ and $\{x \in \mathbb{R} : x^2 - x = 0\}$ are equal, because they have the same elements.

(b) *Empty set*

Everyone with some experience of mathematics knows what is meant by the empty set – a set with no elements. It is an everyday part of mathematics, and will be an essential foundation for our development of abstract set theory and abstract numbers.

(c) *Algebra of sets*

Given any two sets A and B, there are sets $A \cup B, A \cap B, A \backslash B$ (union, intersection and relative complement). These are meaningful irrespective of the nature of the elements of A and B.

The reader is assumed to be familiar with these notions. Also, the more general notions of union and intersection of indexed collections of sets will be familiar. A collection $\{A_i : i \in I\}$ of sets may be given, where I is some index set.

$$\bigcup \{A_i : i \in I\} = \{x : x \in A_i, \text{ for some } i \in I\},$$

and

$$\bigcap \{A_i : i \in I\} = \{x : x \in A_i, \text{ for all } i \in I\}.$$

These can be generalised still further to the case of arbitrary sets whose elements are sets. Given a set X,

$$\bigcup X = \{x : x \in y, \text{ for some } y \in X\},$$

and

$$\bigcap X = \{x : x \in y, \text{ for all } y \in X\}.$$

As with the comprehension principle, we can find trouble with the operations in the algebra of sets, however. Given a set A, it may be thought reasonable to regard $\{x : x \notin A\}$ as meaningful. If we were to denote this by $\bar{A}$, then $A \cup \bar{A}$ would be the set of all sets, and a paradox such as Russell's paradox can be derived from its existence (see Exercise 9 on page 129). *Relative* complements, however, are intuitively clear and trouble free.

(d) *Power set*

This is a familiar notion (introduced in Chapter 2). Given any set, there is a set whose elements are all the subsets of the given set. Again this is meaningful independently of the nature of the objects in the given set.

(e) *Set constructions*

To avoid the difficulties we have noted in regard to the comprehension principle, let us note here the unexceptional ways of constructing or describing sets. First of all, there is no difficulty about collecting a finite number of specified objects into a set. Given any objects $a_1, a_2, \ldots, a_n$ we can form the set $\{a_1, \ldots, a_n\}$. The formation of singleton sets is a special case of this.

Next, we may observe that, whatever apparent need there is for the comprehension principle, in mathematics it is not needed. In any mathematical context, certain objects and sets of objects will be under discussion and it may be convenient to classify the elements of some sets by whether they have a particular property, for example primeness as a property of natural numbers. Thus, given $\mathbb{N}$, there is no problem about the formation of the set $\{x \in \mathbb{N} : x \text{ is prime}\}$. This idea, of forming subsets of given sets through properties which the elements may or may not have, was Zermelo's way of avoiding the contradiction which is a consequence of the comprehension principle. He embodied it in his *separation axiom*, which asserts that this way of constructing sets is legitimate. We shall discuss it in detail in the next section.

There is another way of constructing or specifying sets, and that is by listing the elements. In the case of finite or countable sets it is clear what is meant by this, but there is a more general idea in the background. We may specify a set by giving a function and a domain set for which the set to be constructed is the image set. Such a function 'lists' its image set in a certain sense. For example, consider the function which associates each set x belonging to a domain set A with its power set $P(x)$. Then $\{P(x) : x \in A\}$ is an intuitively reasonable way of specifying a set. It is this procedure which is the essence of the *replacement axiom*. Again, the details are given in the next section.

(f) *Cartesian product*

Given sets A and B, the set $A \times B$ consists of all ordered pairs (x, y) with $x \in A$ and $y \in B$.

All students of mathematics are well aware of this notion, and have a perfectly good understanding of what is mean by an ordered pair. But the awkward set theorist (who has no geometric or algebraic aids like points in a plane or 2-vectors) will still pose the question: given two arbitrary objects a and b, what *is* the ordered pair (a, b)? For formal set theory this requires an answer, and one will be given in Section 4.3.

Of course the idea of Cartesian product leads on to the ideas of relation and function. These are not basic notions of set theory; rather they are mathematical. But they are absolutely essential for mathematics and they fit very easily into the context of set theory. A relation is nothing more than a set of ordered pairs (i.e. a subset of some Cartesian product – see page 83). A function is a relation which is single valued.

▶ Before we proceed to the Zermelo–Fraenkel axioms as such, let us just mention the development of the theory of numbers within set theory.

Very early on, the generality of the idea of set was seen to be such that natural numbers could be represented as sets of a particular kind. Recalling material from Chapter 1, the set of natural numbers is in essence an infinite sequence, starting with 0, each member having a unique successor. The induction principle expressed the most important property of this sequence, and the arithmetic operations had properties demonstrable by that means. Now iteration of an operation is an exceedingly simple idea, so why not start with $\emptyset$ and apply a set-theoretic operation over and over, to generate an infinite sequence? Zermelo took $\emptyset$, $\{\emptyset\}$, $\{\{\emptyset\}\}$, $\{\{\{\emptyset\}\}\}$, ... to represent numbers, by means of the obvious correspondence with $0, 1, 2, 3, \ldots$. It is a non-trivial matter to see that the induction principle can be made to work and to define the arithmetic operations, but it can be done. In this way, set theory can be made to embrace number theory and, consequently, by means of the procedures of Chapter 1, all of the other number systems also. We shall return to this in Section 4.3.

Exercises

1. Translate each of the following into a statement about sets. (The variables are presumed to stand for natural numbers.)
   (i) If x is even then x is not prime. (This is false.)
   (ii) x has a rational square root if and only if x is a perfect square.
   (iii) If x is prime then either $x + 2$ is prime or $x + 4$ is prime. (This is false.)
   (iv) Either x is odd or x is even.
   (v) If 3 divides x and 4 divides x then 12 divides x.
   (vi) There is no prime number x which is a perfect square.
2. The statement '$x = \emptyset$' is equivalent to the statement 'there is no y such that $y \in x$'. Rewrite each of the following without using the symbol $\emptyset$.
   (i) $x \neq \emptyset$.
   (ii) $x \cap y = \emptyset$.
   (iii) If $x \neq \emptyset$ and $y \neq \emptyset$, then $x \cup y \neq \emptyset$.

4.2 The Zermelo–Fraenkel axioms

As we have seen, there is a need for a common starting point, that is to say a list of agreed properties of sets. We shall give such a list, but first let us pause and consider what is meant by an *axiom*. This word is used in different contexts, and we must have a clear idea of what it will mean for us. In one sense, axioms are self-evident truths, and this is the sense in which Greek geometers and nineteenth-century schoolteachers regarded them. In another sense, axioms are just a way of making definitions, particularly in abstract algebra. The axioms for groups, for example, are not unquestionable truths, nor are they intended to be. By definition, a group is a mathematical system in which the group axioms hold.

Axioms of these sorts are different from the sort that we need for set theory. Set theory is different from group theory in the important sense that the notion of set is an intuitively based one, whereas the notion of group is a very precise purely mathematical one. We are not free to define sets in the way that groups are defined, by means of a list of axioms. On the other hand, there is no rigid collection of self-evident truths on which set theory can be based, as Greek geometry was. There are some simple obvious properties, of course, but we shall very soon find in this chapter that some properties of sets in the abstract can be really quite obscure. There are not enough self-evident properties to provide an axiom system which is strong enough. The axiom of choice and the continuum hypothesis are the most celebrated examples of assertions about sets whose acceptability is uncertain. Consequently, we are forced to move away from the conception of axioms as self-evident truths, and to consider set theory axioms as expressing properties of sets which we hope are characteristic and sufficient, and, more importantly, which we *hope* are true.

Too often nowadays books on set theory begin on page 1 with axiom 1 and proceed to describe set theory in a formal way, and to develop the number systems from it. The impression given by such an approach is that the axioms have been handed down on tablets of stone and are consequently indubitable, and that all of mathematics follows from them and is consequently also indubitable. The fact that it is very difficult to learn what formal set theory is about from such an exposition is another unfortunate aspect of such a treatment which is often overlooked.

The axioms for set theory are tentative. Most of them are apparently true. Some of them are currently argued about. Some of them are dismissed as false by some mathematicians. Nobody knows whether they are consistent (i.e. whether any contradictions can be derived from them). They are a starting point. Zermelo initiated a continuing process by enumerating his list in 1908, which has been modified several times since then. Each modification is made in the light of careful consideration of the logical consequences of the currently accepted axioms, or perhaps even in the light of changing intuitions, possibly brought about by such consideration. It should be emphasised that foundations of set theory is a dynamic subject, for set theory comes at the interface of formal mathematics with intuitive ideas, with psychology and the nature of language in the background, and in these areas knowledge can never be complete.

Set theory is about sets in the abstract; so as far as possible particular mathematical ideas must be kept out at this stage. In particular, the properties of sets which we list should be independent of the nature of the elements of the sets. The ingenious way of achieving this is an invention of Fraenkel – all of the objects referred to in the axioms and considered in our formal theory are *sets*. All variables will stand for sets and all elements of sets will be sets. There is no possibility then that properties of other sorts of object will be implicitly assumed. It may be questioned, however, whether this restriction is not so great as to make the set theory useless, since in practice we use properties of sets of numbers, sets of functions, and so on. Perhaps surprisingly, there is no difficulty here, and we shall return to the reasons why in Section 4.3.

Now we come to the axioms themselves. The list which we shall give is essentially that given by Zermelo in 1908, with modifications suggested by Skolem & Fraenkel in 1922. The system of abstract set theory which they determine is called ZF, after the two individuals whose contributions were most significant.

First, let us note that the axioms contain no definition of 'set' or of 'belongs to'. These are undefined primitive notions, whose properties are expressed by the axioms. An abstract set is just something which behaves as the axioms require, and the relation of belonging to (denoted in the usual way by $\in$) is to be governed only by the axioms. We are attempting to make a list which includes or implies all intuitive properties of sets, so we must be careful to avoid making implicit assumptions along the way. Notice, however, that 'set', 'belongs to' and 'equals' are the *only* undefined notions. Other standard notions of set theory such as $\subseteq$, $\cup$, $\emptyset$, etc., are indeed defined by (or as a consequence of) the axioms. There are nine axioms (two of which are axiom schemes with infinitely many instances) in the list which we shall give for ZF, and they are as follows. After the list of the axioms we shall discuss them individually in some detail.

(ZF1) (extensionality)

Two sets are equal if and only if they have the same elements.

(ZF2) (null set)

There is a set with no elements.

(ZF3) (pairing)
Given any sets x and y, there is a set whose elements are x and y.

(ZF4) (union)
Given any set x (whose elements are sets), there is a set which has as its elements all elements of elements of x.

(ZF5) (power set)
Given any set x, there is a set which has as its elements all subsets of x.

(ZF6) (separation)
Given any well-formed formula $\mathscr{A}(y)$ and any set x, there is a set $\{y \in x : \mathscr{A}(y)\}$.

(ZF7) (replacement)
Given any well-formed formula $\mathscr{F}(x, y)$ which determines a function and any set u, there is a set v consisting of all objects y for which there is $x \in u$ such that $\mathscr{F}(x, y)$ holds.

(ZF8) (infinity)
There is a set x such that $\emptyset \in x$, and such that, for every set $u \in x$, we have $u \cup \{u\} \in x$.

(ZF9) (foundation)
Every non-empty set x contains an element which is disjoint from x.

Axiom (ZF1) (extensionality axiom)

Two sets are equal if and only if they have the same elements.

In more formal terms, a necessary and sufficient condition for two sets x and y to be equal is that for every set z, $z \in x \Leftrightarrow z \in y$.

This principle expresses the fundamental property that a set is determined by its elements.

Another way of writing (ZF1) is to express the necessary and sufficient condition for equality of x and y as:

for every set z, $(z \in x \Rightarrow z \in y)$ and $(z \in y \Rightarrow z \in x)$.

Now anyone with any experience of sets at all would write this last in the form

$$(x \subseteq y) \quad \text{and} \quad (y \subseteq x),$$

and this leads us to introduce into our formal theory the symbol $\subseteq$ by means of a definition, in terms of the primitive notion $\in$.

Definition

For sets x and y, *x is a subset of y* means: for every set z, $z \in x \Rightarrow z \in y$.

► (ZF1) can therefore be rewritten: for sets x and y, $x = y$ if and only if $x \subseteq y$ and $y \subseteq x$.

Axiom (ZF2) (null set axiom)

There is a set with no elements.

It is an exercise in trivial logical manipulation to show, using (ZF1), that there is only one set with no elements (two such sets must be equal). We use the standard symbol $\emptyset$ for the null set, i.e. the empty set. This axiom is included in order to ease understanding, but it in fact is a consequence of the other axioms. (See Exercise 20 on page 129.)

Axiom (ZF3) (pairing axiom)

Given any sets x and y there is a set u whose elements are x and y.

This corresponds to one of the procedures for constructing sets described earlier, but restricted to pairs of sets. However, (ZF3) includes singletons; if we take the special case where $x = y$, we obtain the assertion:

Given any set x there is a set u whose only element is x.

The normal notations for pairs and singletons are $\{x, y\}$ and $\{x\}$ respectively. Notice that the pair here is an *unordered* pair. (ZF1) easily implies that $\{x, y\} = \{y, x\}$, for any sets x and y.

Axiom (ZF4) (union axiom)

Given any set x (whose elements are sets), there is a set which has as its elements all elements of elements of x.

The standard notation, introduced in the previous section, is $\bigcup x$. We have $z \in \bigcup x$ if and only if there is a set $y \in x$ for which $z \in y$.

How can we recover from this the notion of the union of two sets? Given two sets x and y, $x \cup y$ is to consist of all elements of x, together with all elements of y. Thus we first form the set $\{x, y\}$, according to (ZF3) and then form $\bigcup \{x, y\}$, to obtain a set with the appropriate elements. Axioms (ZF3) and (ZF4), together, then guarantee that unions of two sets can be formed.

Definition

$x \cup y$ means $\bigcup \{x, y\}$, for sets x and y.

▶ The commutative and associative laws can be proved for $\cup$, using (ZF1). These are left as exercises; notice that the commutative property follows from the equality $\{x, y\} = \{y, x\}$, which has already been noted.

Axiom (ZF4) is stated in terms of $\bigcup$ rather than $\cup$ because it has wider scope than just unions of pairs of sets. Clearly, it would have been possible to make the axiom more intelligible by splitting it into two and giving the case of the union of two sets as an axiom. There are two conflicting purposes here. One is ease of understanding of individual axioms. The other is drawing a line somewhere to prevent unnecessary proliferation of axioms. The system as a whole will be easier to comprehend if the number of axioms is small.

In the background, though not necessarily significant for us, is the desirability of avoiding redundant axioms. The mathematician has no need to state as an assumption an assertion which follows from other explicitly stated assumptions.

Bearing this in mind, let us observe that we do not need an axiom now to cover the construction of finite sets with specified elements which was mentioned in the previous section. Singletons, pairs and unions, as given in (ZF3) and (ZF4), will do it for us. First, for *triples*: let x, y and z be sets, and define $\{x, y, z\}$ to be $\bigcup \{\{x, y\}, \{z\}\}$. Inductively, for $n \geqslant 3$, we can define (for sets $x_1, x_2, \ldots, x_n$) $\{x_1, \ldots, x_n\}$ to be $\bigcup \{\{x_1, \ldots, x_{n-1}\}, \{x_n\}\}$.

It would seem to make sense to include as our next axiom one which covers intersections of sets. This is also a case, however, where other axioms make such an axiom unnecessary. The existence of intersections will be an easy consequence of Axiom (ZF6) which follows, as we shall see shortly.

Axiom (ZF5) (power set axiom)

Given any set x, there is a set which has as elements all subsets of x.

This is a standard notion with a standard notation: $P(x)$ stands for the power set of x.

▶ This ends the first group of axioms – those which correspond to basic set properties and constructions. The remainder are rather more subtle, and though they are apparently intuitive truths, we should be rather more careful about their significance.

Axioms (ZF6) and (ZF7) are related to the comprehension principle. This is the part of Zermelo's original system which proved unsatisfactory and which led to the modifications by Skolem & Fraenkel. Zermelo postulated the separation axiom: given any set x and any 'definite property' P, there is a set consisting of all those elements of x which have property P. This reflects the process which we have already mentioned of 'separating out' a subset of a given set. The difficulty lies with the notion of definite property, which Zermelo did not explain very well. Skolem's contribution was to make this precise. He required the assertion 'x has property P' to be expressible by means of a formula built up from propositions of the form $a \in b$ or $a = b$ (where a and b represent arbitrary variables standing for sets), using logical operations such as conjunction, disjunction, negation and implication, and using quantifiers in the normal way. This idea of formula is taken from mathematical logic, and although we shall not be dealing with any of its logical aspects, it will be important for the reader to have a grasp of what is meant by it, as we shall use it again several times. A formula such as the above we shall refer to as a *well-formed formula* of ZF.

We shall use standard symbols as follows.

conjunction:	$P \& Q$ stands for P and Q,
disjunction:	$P \lor Q$ stands for P or Q (or both),
implication:	$P \Rightarrow Q$ stands for P implies Q,
negation:	$\neg P$ stands for not P,
universal quantifier:	$(\forall u)$ means 'for all sets u',
existential quantifier:	$(\exists u)$ means 'there is a set u such that'.

Let us consider some examples.

Examples 4.4

Let x and y be fixed sets.

(a) $\{z \in x : z \in y\}$ fits the pattern of the separation axiom, where the property concerned is expressed by the formula $z \in y$. Note that this set is just $x \cap y$.

(b) $\{z \in x : z$ is a subset of $y\}$ fits the pattern of the separation axiom. To see this, we must find a well-formed formula of ZF which expresses the property 'z is a subset of y'. As we have recently noted, the formula $(\forall u)(u \in z \Rightarrow u \in y)$ does just this. Notice that $z \subseteq y$ is *not* a well-formed formula of ZF, since it contains the defined symbol $\subseteq$.

(c) $\{z \in x : z \cap y = \emptyset\}$. Here the formula required is $(\forall u)(\neg(u \in z \ \& \ u \in y))$. Again, $z \cap y = \emptyset$ is not a well-formed formula of ZF, but it is equivalent to the formula given above, which is.

(d) Notice that the same set may be written in different ways, using different formulas. For example $\{z \in x : z = z\}$ and $\{z \in x : z \subseteq z\}$ are the same set.

▶ Let us now state the axiom.

Axiom (ZF6) (separation axiom)

Given any well-formed formula $\mathscr{A}(y)$ of ZF (expressing an assertion about the set y), and given any set x, there is a set whose elements are all those elements y of x for which $\mathscr{A}(y)$ holds. More formally, there is a set $\{y \in x : \mathscr{A}(y)\}$.

It is important to note that this is not a single axiom. It is what is usually termed an axiom *scheme*. There are infinitely many possible formulas $\mathscr{A}(y)$, so there are infinitely many instances of (ZF6).

Remarks 4.5

(a) As noted above, given any sets x and y, there is a set $x \cap y$, using (ZF6). More generally, let x be any set (whose elements are sets).

$$\bigcap x = \{y \in \bigcup x : (\forall u)(u \in x \Rightarrow y \in u)\}.$$

That is, $\bigcap x$ is the set of all elements y which belong to every element u of x. All such y must belong to $\bigcup x$, so this is the set of which $\bigcap x$ is to be specified as a subset, using (ZF6).

(b) Let x and y be given sets. Using (ZF6), we may define the relative complement $x \backslash y$ as follows.

$$x \backslash y = \{z \in x : z \notin y\}.$$

▶ It may seem restrictive to allow only formulas involving $\in$, $=$ and logical operations, and in one sense it is. But what alternative have we? We must not use any notions that bring with them any implicit assumptions, since these will affect our 'basic' properties of sets. We must therefore use only $\in$ and $=$ in the formulas, these being the only basic relations in our theory. As shown in Examples 4.4, it is possible to allow also symbols like $\subseteq$ and $\cap$ in the formulas, since such formulas are equivalent to formulas involving only $\in$ and $=$.

There is another comment which should be made about (ZF6), and that is that it is *impredicative*. What this means is that the formula $\mathscr{A}(y)$ used to specify the subset in question might possibly contain a universal quantifier, and so for $\mathscr{A}(y)$ to be satisfied we might require something to be true for *all* sets, including the one which is being specified by $\mathscr{A}(y)$. On the face of it there is a vicious circle here, and indeed this is a substantial difficulty. Of course, not all formulas would lead to this trouble. We give some examples.

Examples 4.6
The following formations of sets are impredicative.
(a) $\{y \in x : (\forall u)(u = u)\}$.
(b) $\{y \in x : (\forall u)(u \in y \Rightarrow u \in x)\}$.
(c) $\{y \in x : (\forall u)(u \subseteq x \Rightarrow u \leqslant y)\}$.

Note that the specifying formula in (c) is not a well-formed formula of ZF. An equivalent formula can be found which is well-formed, using the methods of Section 4.3. The impredicativity in (a) is entirely artificial. This set can be defined in other ways which are unobjectionable, for example as $\{y \in x : y = y\}$. The set (b) is not as artificial, but nevertheless the impredicativity can be removed. This set is just $x \cap P(x)$, and may be specified as $\{y \in P(x) : y \in x\}$. The case of (c) is more difficult to resolve. The reader may care to ponder whether the impredicativity here can be removed.

▶ Most workers in foundations of mathematics (the intuitionists are a notable exception) are prepared to accept (ZF6) as it is, in spite of the unsatisfactory possibility of sets being defined impredicatively. While no contradictions are known to follow from it, there is a difficulty in attaching meaning to objects defined by apparently circular formal definitions. Various attempts have been made to improve the situation, but none has turned out entirely satisfactorily.

Axiom (ZF6) is an expression of a legitimate way of collecting together objects into one set. That there are ways which are not legitimate is demonstrated by Russell's paradox. Consequently, our axioms must give expression (we hope) to all ways that are legitimate. Zermelo's axiom (ZF6) was accepted for a time as the best guess. However, it became clear to Fraenkel in 1922 (and others) that (ZF6) was not adequate to cover all intuitive set constructions. This is the background to Axiom (ZF7), which Fraenkel postulated in order to meet his difficulty. Here is the difficulty:

Example 4.7

Let x be a fixed set. Then $P(x)$ is given by (ZF5), and $P(P(x))$ likewise, and so on. We can generate a sequence of sets $x, P(x), P(P(x)), \ldots$. It is intuitively sensible to regard the collection of all of these as a set, and to denote it by

$$\{x, P(x), P(P(x)), \ldots\}.$$

There is no way of showing from Zermelo's axioms that there is a set with these elements. We cannot go into a demonstration of this, but the underlying difficulty is that (ZF6) cannot be made to apply in this situation because the set in question is not constructed as a *subset* of a given set.

▶ More generally, the inadequacy in Zermelo's axioms was that they do not allow for the construction of sets whose elements are listed by a function or rule. Axiom (ZF7) is designed for this purpose.

Axiom (ZF7) (replacement axiom)

Let $\mathcal{F}(x, y)$ be a well-formed formula of ZF, which determines a function. Then, given any set u, there is a set v consisting of all images of elements of u under the function, i.e. there is a set v consisting of all objects y for which there is $x \in u$ such that $\mathcal{F}(x, y)$ holds.

Another less formal way of stating this is:
The image of a set under any function is a set.

Again, we should observe that (ZF7) is an axiom *scheme*, just as (ZF6) was, with infinitely many instances, one for each formula $\mathscr{F}(x, y)$.

Shortly we shall explain the terms used in this axiom, but first let us see how (ZF7) applies to our problematic Example 4.7. The collection $\{x, P(x), P(P(x)), \ldots\}$ is the image of the set of natural numbers under the function which associates 0 with x and associates n (>0) with $P^n(x)$. We shall discuss later the details of how the set of natural numbers arises in formal set theory, but for the moment we may assume that it is indeed a set, just in order to make sense of this example.

Now what does it mean for a formula $\mathscr{F}(x, y)$ to determine a function? Axiom (ZF6) used the idea of a formula built up from propositions of the form $a \in b$ and $a = b$ using logical operations. In that case the formulas were all assumed to represent assertions $\mathscr{A}(y)$ about the unspecified object y. In the present context we use the idea of a formula constructed in the same way but with two 'free' variables. Examples of such formulas are $(\forall u)(u \in x \Rightarrow u \in y)$, $(\forall u)(u \in x \Rightarrow (\exists v)(u \in v \ \& \ v \in y))$, where in each case the free variables are x and y.

Definition
A well-formed formula $\mathscr{F}(x, y)$ with two free variables *determines a function* if for any sets x, y and z, $\mathscr{F}(x, y)$ and $\mathscr{F}(x, z)$ imply $y = z$. Equivalently, given any set x there is at most one set y such that $\mathscr{F}(x, y)$ holds.

Examples 4.8
(a) The formulas given above: $(\forall u)(u \in x \Rightarrow u \in y)$ and $(\forall u)$ $(u \in x \Rightarrow (\exists v)(u \in v \ \& \ v \in y))$ do *not* determine functions.
(b) The formula $y = x$ determines a function, in a trivial way.
(c) The formula $x = x \ \& \ (\forall u)(u \notin y)$ determines a function (intuitively, the function with value $\emptyset$ everywhere).
(d) The formula $(\forall u)(u \in y \Leftrightarrow (\exists v)(u \in v \ \& \ v \in x))$ determines a function. Notice that in this case the formula is equivalent to: $y = \bigcup x$.

▶ Perhaps surprisingly, this new axiom of Fraenkel's which closes a gap left in Zermelo's system, is related to (ZF6), in the following sense.

Theorem 4.9
Axiom scheme (ZF7) implies Axiom scheme (ZF6).

Proof

Let $\mathscr{A}(y)$ be a formula as specified in (ZF6), and let x be a given set. We may regard $\mathscr{A}(y)$ as expressing a condition to be satisfied or not by elements of x. Let $\mathscr{F}(u, v)$ be the formula:

$$u = v \ \& \ \mathscr{A}(u).$$

Intuitively this formula determines the identity function on the collection of all u for which $\mathscr{A}(u)$ holds. Restricting this function to x, it becomes the identity function on the collection of all elements u of x such that $\mathscr{A}(u)$ holds. The set

$$w = \{v : (\exists u)(u \in x \ \& \ \mathscr{F}(u, v))\}$$

is given by (ZF7) and is just the same as the set

$$\{v \in x : \mathscr{A}(v)\}$$

required by (ZF6).

▶ (ZF6) is thus a redundant axiom (scheme). It is not necessary to have it as part of the basic common starting point. It is normally included among the Zermelo–Fraenkel axioms, however, because it represents the standard way of constructing sets.

The eighth axiom serves a very specific purpose. The first seven axioms yield the standard properties concerning membership, unions, intersections, and the normal ways of constructing sets. (ZF7) even gives an explicit procedure for constructing an infinite set, as the example of $\{x, P(x), P(P(x)), \ldots\}$ shows. But notice that that construction depended on the notion of the set of natural numbers, and in general (ZF7) cannot yield a construction of an infinite set without reference to a previously constructed infinite set. Unions and power sets likewise do not lead from finite to infinite sets. Since our system would clearly be inadequate if there were no infinite sets mentioned by it, how are we to embody the possibility of constructing infinite sets among the axioms? Axiom (ZF8) does the job.

Axiom (ZF8) (infinity axiom)

There is a set x such that $\emptyset \in x$, and such that for every set $u \in x$ we have $u \cup \{u\} \in x$ also.

This axiom does more than assert that there is a set which is infinite. It gives an explicit description of the elements it must contain, namely

$$\emptyset, \emptyset \cup \{\emptyset\}, \emptyset \cup \{\emptyset\} \cup \{\emptyset \cup \{\emptyset\}\}, \ldots.$$

Notice that the set in question may have other elements also. We shall see in the next section that it follows from (ZF8) that the members of the sequence above constitute a set by themselves. It is apparent, although a formal proof is not particularly easy, that the members of this sequence are all distinct, which of course is what makes the set infinite. Now why do we choose this complicated sequence? Why not choose the sequence mentioned earlier, namely $\emptyset, \{\emptyset\}, \{\{\emptyset\}\}, \dots$, which Zermelo used to construct the natural number system? The answer is that we could have chosen this sequence, or indeed others. The choice of $\emptyset, \emptyset \cup \{\emptyset\}, \emptyset \cup \{\emptyset\} \cup \{\emptyset \cup \{\emptyset\}\}, \dots$ follows work of von Neumann (1923). It facilitates two developments which will be covered in subsequent sections. The first is the construction of the natural numbers within set theory, and the second is the convenient and useful definition of ordinal number. These matters will be discussed in Section 4.3 and Chapter 6.

The final axiom in the system ZF is a technical one, in the sense that it is not essential for the development of standard mathematics from set theory. It is included primarily to reflect the following intuitive property of sets. Starting with any set x_0, choose an element x_1 of x_0, choose an element x_2 of x_1, choose an element x_3 of x_2, and so on. Intuition suggests that this process ought to stop, i.e. that there does not exist an infinite sequence $x_0, x_1, x_2, \dots$ of sets such that $x_{n+1} \in x_n$ for each n. A particular case of this is the impossibility of a set being an element of itself, for if $x \in x$ then $x, x, x, \dots$ is an infinite sequence of the kind referred to above.

Axiom (ZF9) (foundation axiom)

Every non-empty set x contains an element which is disjoint from x.

Let us convince ourselves that (ZF9) does serve the purpose described above. Let x be a set with elements $x_0, x_1, x_2, \dots$, and suppose that $x_{n+1} \in x_n$ for each $n \in \mathbb{N}$. By (ZF9), x contains an element y for which $y \cap x = \emptyset$. But we must have $y = x_k$ for some $k \in \mathbb{N}$, and certainly $x_{k+1} \in x_k$ and $x_{k+1} \in x$, so that $x_{k+1} \in x \cap y$. This contradiction yields the conclusion that (ZF9) implies the non-existence of such sequences and, consequently, the impossibility of $x \in x$.

Example 4.10

An application of (ZF9) is to demonstrate that for any set x, $x \cup \{x\} \neq x$. If we have $x \cup \{x\} = x$ it would immediately follow that $x \in x$, which we know to be impossible. So here is a process for constructing a sequence of distinct sets, starting from any given set x:

$$x, x \cup \{x\}, x \cup \{x\} \cup \{x \cup \{x\}\}, \dots .$$

An example of this occurs in Axiom (ZF8), where x was taken to be $\emptyset$ (it should be noted in passing that in that particular case, (ZF9) is not required for the proof that the members of the sequence are distinct). This construction process is also essential to the notion of ordinal number, as discussed in Chapter 6.

▶ Before closing this section let us not forget the axiom of choice, which has been mentioned in Chapter 3. Although it was included in Zermelo's original list in 1908 (Zermelo was the first to state it explicitly as an axiom), nowadays it is customary not to include it as one of the axioms of ZF. This is perhaps because mathematicians have over the years treated the axiom of choice with some suspicion and perhaps because (more recently) researchers have been more and more interested in set theory without it. We shall discuss the reasons for these attitudes in the next chapter. Here we give a formal statement of the axiom.

(AC) (axiom of choice)

Given any (non-empty) set x whose elements are pairwise disjoint non-empty sets, there is a set which contains precisely one element from each set belonging to x.

The system of set theory determined by Axioms (ZF1) to (ZF9) and including (AC) also is usually denoted by ZFC.

Exercises

1. Using (ZF1), prove that the null set is unique. Prove also that $\emptyset \subseteq x$ holds for every set x.
2. Using (ZF1), prove that $\{x, y\} = \{y, x\}$ for all sets x and y.
3. Let $x = \{\{\{y\}\}, \{\{y, \{z\}\}\}\}$. Find the elements of $\bigcup x$, $\bigcup\bigcup x$, $\bigcup\bigcup\bigcup x$.
4. Prove the following, for sets x, y, z:
   (i) $x \cup y = y \cup x$.
   (ii) $x \cup (y \cup z) = (x \cup y) \cup z$.
   (iii) $x \cap y = y \cap x$.
   (iv) $\bigcup\{x, y, z\} = (x \cup y) \cup z$.
5. Prove that, for any sets x and y, if $\bigcup x \neq \bigcup y$ then $x \neq y$. Is it the case that $\bigcup x = \bigcup y$ implies $x = y$?
6. Find the elements of $P(P(P(P(\emptyset))))$.
7. Show that $\bigcup P(x) = x$ for every set x. Show also that $\bigcap P(x) = \emptyset$. What can be said about $P(\bigcup x)$ in general?

8. Using (ZF5) and (ZF6) but *not* (ZF3), show that for any set x there is a set which has x as its sole element.

9. Using (ZF6), derive a contradiction from the supposition that there is a set U of all sets. (Hint: use the formula $x \notin x$, and obtain a contradiction as in Russell's paradox.)

10. Let x be a fixed non-empty set. Derive a contradiction from the supposition that $\{y : y \sim x\}$ is a set (here $\sim$ means cardinal equivalence, as in Chapter 2). (Hint: start by showing that for any set z there is a set y with $z \in y$ such that $y \sim x$; it will then follow that $\bigcup \{y : y \sim x\}$ contains every set, and a contradiction comes as in Exercise 9 above.)

11. (i) Derive a contradiction from the supposition that $\{H : H$ is a group and H is isomorphic to $G\}$ is a set, where G is some fixed group.
    (ii) Derive a contradiction from the supposition that $\{y : x \subseteq y\}$ is a set, where x is some fixed set.

12. Can we define $\{y \in x : (\forall u)(u \subseteq x \Rightarrow u \preccurlyeq y)\}$ avoiding the impredicativity?

13. Given a set x, justify (using the axioms of ZF) the existence of the set $\{\{a\} : a \in x\}$.

14. Deduce (ZF3) as a consequence of (ZF8) and (ZF7). (Hint: find an appropriate formula determining a function such that the required unordered pair is the image under that function of the set given by (ZF8).)

15. Consider the following weak version of (ZF3). Given any sets x and y, there is a set u such that $x \in u$ and $y \in u$. Call this (ZF3′). Prove that (ZF3′) and (ZF6) together imply (ZF3).

16. Write down similar weak versions of (ZF4) and (ZF5), and show that, together with (ZF6), each implies the corresponding strong version.

17. (ZF9) ensures that there is no infinite sequence $x_0, x_1, x_2, \ldots$ with $x_{n+1} \in x_n$ for each n. Give an example of an infinite sequence $y_0, y_1, y_2, \ldots$ of sets such that $y_{n+1} \subseteq y_n$ for each n.

18. Can there exist sets x and y with $x \in y$ and $y \in x$? Which of our axioms are relevant? Is it possible for $x \in y$, $y \in z$, $z \in x$ to hold simultaneously? Generalise.

19. Let us call a sequence $x_0, x_1, x_2, \ldots$ of sets for which $x_{n+1} \in x_n$ for each n a descending $\in$-sequence. (ZF9) implies that every such sequence is finite. Describe a set which contains descending $\in$-sequences of arbitrary length.

20. Write down a statement which is equivalent to (ZF8) but which does not include any occurrence of the symbol $\emptyset$. Show that this alternative infinity axiom together with (ZF6), implies (ZF2).

4.3 **Mathematics in** ZF

The language of ZF is excessively restricted. Originally it allowed only variables, $=$ and $\in$ in addition to the punctuation and logical symbols. We introduced some of the standard mathematical usages, like $\emptyset$, $\subseteq$, $\cap$, $\cup$, etc. in the last section. Here we continue the development of standard mathematical notations, usages and concepts within the framework of our formal system. Although we shall apparently *define* such notions as ordered pair, function and natural number in terms of our formal context, the reader should remember that the object of the exercise is to demonstrate that the system ZF is 'adequate' for mathematics, that is to say that the standard notions and procedures of mathematics can be dealt with in the system. Thus our definition of natural number which follows (for example) should be seen as the definition of 'abstract natural number', which in no way supersedes the clear intuitive knowledge that we all have about natural numbers. The ideas of this section have to do with validating the system ZF as a true reflection of mathematics. They do not re-define familiar notions.

The most important mathematical idea which is not explicitly referred to in ZF is that of a relation (a function is of course a particular kind of relation). We have previously considered a relation to be a *set* of ordered pairs, and this is how we fit the idea into ZF. But first we must think about ordered pairs, which as yet have no place in ZF. The definition may look odd at first but, as we shall see, once the basic properties have been derived, the formalities of the definition can be forgotten.

Definition
For sets x and y, the set $\{\{x\}, \{x, y\}\}$ is called the *ordered pair* of x and y, and denoted by (x, y).

▶ What makes this work? The property that we require is that if $(a, b) = (x, y)$ then $a = x$ and $b = y$ (for any sets a, b, x, y). Let us demonstrate this. Let $(a, b) = (x, y)$, i.e.

$$\{\{a\}, \{a, b\}\} = \{\{x\}, \{x, y\}\}.$$

Hence,

$$\{a\} = \{x\} \quad \text{or} \quad \{a\} = \{x, y\},$$

and

$$\{a, b\} = \{x\} \quad \text{or} \quad \{a, b\} = \{x, y\}.$$

Now $\{a\} = \{x\}$ implies $a = x$, and $\{a\} = \{x, y\}$ implies $a = x = y$ (using (ZF1)), so in any case we have $a = x$. Further, $\{a, b\} = \{x\}$ implies $a = b = x$, so that $(a, b) = \{\{a\}\}$, and since $\{x, y\} \in (a, b)$ we have $\{x, y\} = \{a\}$, and consequently $x = y = a$ ($= b$) and the result follows in this case. Lastly, we have the case where $\{a, b\} = \{x, y\}$ with $a \neq b$. Here $b \in \{x, y\}$, so $b = x$ or $b = y$. But $b \neq x$ since $a = x$ and $a \neq b$. Thus $b = y$ as required.

This definition of ordered pair can be extended inductively to ordered triples, quadruples, etc.

Definition
(i) (x, y, z) denotes $((x, y), z)$ $(x, y, z$ sets).
(ii) For any natural number n, if $x_1, \ldots, x_{n+1}$ are sets, we define $(x_1, \ldots, x_{n+1})$ to be $((x_1, \ldots, x_n), x_{n+1})$.

► There is no particular significance to these definitions except that they work. No properties of ordered pairs will be used except that demonstrated above. The point is that (x, y) can now be regarded as a part of our formal system, standing for an object with certain properties.

The Cartesian product of two sets, x and y, we know to be just the set of all ordered pairs (a, b) with $a \in x$ and $b \in y$. To fit this into ZF requires a little ingenuity. We shall use the separation axiom to construct $x \times y$ within ZF. If $a \in x$ and $b \in y$, then since $(a, b) = \{\{a\}, \{a, b\}\}$ is a subset of $P(\{a, b\})$, and $\{a, b\}$ is a subset of $x \cup y$, we have $(a, b) \subseteq P(x \cup y)$, and so $(a, b) \in P(P(x \cup y))$.

Definition

$$x \times y = \{z \in P(P(x \cup y)) : (\exists a)(\exists b)(a \in x \ \& \ b \in y \ \& \ z = (a, b))\}.$$

The ingenuity is directed towards obtaining a set (in this case $P(P(x \cup y))$) of which the set to be defined is to be a subset, so that (ZF6) can be applied.

This definition can be extended inductively to an arbitrary finite Cartesian product as follows. For any natural number n, if $x_1, \ldots, x_{n+1}$ are sets then $x_1 \times \cdots \times x_{n+1} = (x_1 \times \cdots \times x_n) \times x_{n+1}$.

Definition
(i) A *binary relation* is a subset of a Cartesian product of two sets. That is: $(\exists x)(\exists y)(z \subseteq x \times y)$ means that z is a binary relation. Likewise, n-ary relations can be defined for each natural number n. A binary relation *on a set* x is a subset of $x \times x$.
(ii) The *domain* of a binary relation z is the set of first elements of ordered pairs in z. This is the set $\{x \in \bigcup(\bigcup z) : (\exists y) ((x, y) \in z)\}$. The trick again is to find the set (in this case $\bigcup(\bigcup z)$) of which the required set is a subset, in order to apply (ZF6). The *image* is similarly defined as the set of second elements of ordered pairs.

▶ In Chapter 3 we noted that certain kinds of relation are particularly significant. The conditions in the definitions may be expressed by well-formed formulas of ZF.

Definition
f is a function may be expressed by:

$$(\exists x)(\exists y)(f \subseteq x \times y)\ \&\ (\forall u)(\forall v)(\forall w)(((u, v) \in f\ \&\ (u, w) \in f) \Rightarrow (v = w)).$$

This says: f is a binary relation and f is single valued.

▶ It is left as an exercise for the reader to see how 'f is an injection', 'f is a surjection' and 'f is a bijection' can be expressed by well-formed formulas.

In Chapter 1 we referred to the set of all functions from one set to another. This can be constructed in ZF. Again, we apply (ZF6). If x and y are sets and f is any function whose domain is x and whose image is a subset of y, then f is a subset of $x \times y$, i.e. $f \in P(x \times y)$, and so the set of all such functions is

$$\{f \in P(x \times y) : (\forall u)(u \in x \Rightarrow (\exists v)(v \in y\ \&\ (u, v) \in f)\ \&$$

$$(\forall u)(\forall v)(\forall w)(((u, v) \in f\ \&\ (u, w) \in f) \Rightarrow (v = w))\}.$$

Translated, this says:

$$\{f \in P(x \times y) : \text{domain } f = x \text{ and } f \text{ is single valued}\}.$$

Using the idea of function, we can now fit the notion of an indexed collection (or family) of sets into our system by means of the following definition.

Definition
A *family* of sets is a function F from an index set I to some range set. Intuitively we consider $\{F(i) : i \in I\}$ to be the family of sets.

▶ It should be noted that a family is distinct from a set of sets, since a set may be repeated in a family, but can count only once as an element of a set. For example, the function F from $\mathbb{N}$ to $P(\mathbb{R})$, such that $F(n) = \mathbb{R}$ for every $n \in \mathbb{N}$, is a family, but $\{F(n) : n \in \mathbb{N}\}$ as a *set* has precisely one element.

We identify the indexed collection with the function which does the indexing. This means that care is required over unions, for example. If F is a family of sets, the union of the sets in the family is the union of the image of F, which is not the same thing as $\bigcup F$.

Next we come to another fundamental notion of pure mathematics.

Definition

z *is an equivalence relation on* x may be expressed by:

$(z \subseteq x \times x)$ & $(\forall u)(u \in x \Rightarrow (u, u) \in z)$ & $(\forall u)(\forall v)((u, v) \in z \Rightarrow$

$(v, u) \in z)$ & $(\forall u)(\forall v)(\forall w)(((u, v) \in z$ & $(v, w) \in z) \Rightarrow (u, w) \in z)$.

This just says that z is a binary relation on x, and z is reflexive, symmetric and transitive.

▶ It should be noted at this point that what we have is the notion of equivalence relation *on a set*, and we should observe that equinumerosity of sets is not an equivalence relation in this sense, since there is no appropriate set x. A contradiction similar to Russell's paradox would follow if we assumed otherwise. By Exercise 10 on page 129 we know that for any set x, $\{x : x \sim y\}$ cannot be a set. However, if z is an equivalence relation on a set x then *equivalence classes* can be properly defined as sets. Given an element $u \in x$, the set $\{v \in x : (u, v) \in z\}$ is the equivalence class determined by u.

Definition

z *is an order relation on a set* x is expressed by:

$(z \subseteq x \times x)$ & $(\forall u)(u \in x \Rightarrow (u, u) \in z)$ &

$(\forall u)(\forall v)(((u, v) \in z$ & $(v, u) \in z) \Rightarrow (u = v))$ &

$(\forall u)(\forall v)(\forall w)(((u, v) \in z$ & $(v, w) \in z) \Rightarrow (u, w) \in z)$.

The reader is left with the exercise of finding formulas which express the properties of being a total order and being a well-order.

▶ We see from the above examples how mathematical objects at first sight distinct from sets can be defined within set theory as sets of particular kinds. So far we have dealt with the basic notions of Chapters 2 and 3 and these provide the tools for a good deal of abstract algebra. However what we have not yet mentioned is perhaps a harder task, and that is to develop and define numbers within our set theory.

Let us first give an intuitive basis for the definition we shall give of the set of abstract natural numbers. Given any set x let us denote by x^+ the set $x \cup \{x\}$. We have noted that by (ZF9), $x \cup \{x\}$ is always distinct from x. x^+ is called the *successor* of x.

Using the successor operation, we can construct a sequence of sets:

$$\emptyset, \emptyset^+, \emptyset^{++}, \emptyset^{+++}, \ldots .$$

Abstract natural numbers will be the terms in this sequence of sets, and we shall use boldface type to denote these numbers. Let us write down a few explicitly.

0 is $\emptyset$.
1 is $\emptyset^+$, i.e. $\{\emptyset\}$.
2 is $\emptyset^{++}$, i.e. $\{\emptyset, \{\emptyset\}\}$.
3 is $\emptyset^{+++}$, i.e. $\{\emptyset, \{\emptyset\}, \{\emptyset, \{\emptyset\}\}\}$.

The reader should check that these sets actually have the elements listed. Observe that **0** is a set with no elements, **1** is a set with one element, **2** has two elements and **3** has three elements. It is easy to see in general that if x is a set with finitely many elements then $x \cup \{x\}$ has one additional element. By an informal inductive argument we can see that each abstract natural number n is a set with n elements. This property of the abstract natural numbers as listed above is one of the main reasons for choosing to define them in this way rather than in the way that was mentioned earlier, using the sequence $\emptyset, \{\emptyset\}, \{\{\emptyset\}\}, \ldots .$

The objects which we have labelled **0, 1, 2,** ... are sets of a particular kind. Before we go any further, what we would like to know is that the collection of all of them, namely, $\{\textbf{0}, \textbf{1}, \textbf{2}, \ldots\}$, is a legitimate object of discussion with ZF. This is where the infinity axiom (ZF8) comes in. As observed before, without (ZF8) there is no procedure in ZF for assembling together into a set any infinite collection of objects. (ZF8) itself is in fact specifically designed so as to guarantee the legitimacy in ZF of the *set* $\{\textbf{0}, \textbf{1}, \textbf{2}, \ldots\}$. Once this has been established, of course, other axioms (power set, separation, replacement) may be used as we saw in Section 4.2 to construct other infinite sets.

Definition

A set x is said to be a *successor set* (or *inductive set*) if $\emptyset \in x$ and if, for each $y \in x$, we have $y^+ \in x$ also.

▶ Notice two things at this stage. First, the infinity axiom asserts that there is a successor set and, second, in any successor set the members of the sequence $\emptyset, \emptyset^+, \emptyset^{++}, \ldots$ must all be elements. What we are seeking is a successor set which has no other elements.

Theorem 4.11

There is a minimal successor set, i.e. a successor set which is a subset of every other successor set.

Proof

By (ZF8), there is a successor set x, say. By (ZF5) and (ZF6), there is a set

$$v = \{u \in P(x) : u \text{ is a successor set}\}.$$

Then v is the collection of all subsets of x which are successor sets. A successor set which is a subset of every other successor set must in particular be a subset of x (if it exists at all). $\bigcap v$ is in fact the set we need. It is easy to show that any intersection of successor sets is a successor set (an exercise for the reader), so $\bigcap v$ is a successor set. To see that it is minimal, let z be any successor set. Then $x \cap z$ is a successor set, by the above result about intersections, and $x \cap z \subseteq x$, so $x \cap z \in v$. Now $\bigcap v$ is necessarily a subset of every element of v, so $\bigcap v \subseteq x \cap z$, and hence $\bigcap v \subseteq z$ as required.

▶ Now we can give our formal definition of natural numbers.

Definition

Abstract natural numbers are the elements of the minimal successor set as given by Theorem 4.11. This minimal successor set is denoted by ω. Of course, since ω is a successor set, the members of the sequence $\emptyset, \emptyset^+, \emptyset^{++}, \ldots$ are all elements of ω. Putting this another way, **0, 1, 2**, ... as defined above are all abstract natural numbers.

▶ This somewhat roundabout procedure was necessary in order to avoid intuitive processes which are not part of ZF. We have not explicitly shown that the elements of ω are **0, 1, 2, 3**, ... and nothing else. Using (ZF8) we work downwards to the set ω, so to speak, since the more natural process of building up sets from their members is not available here. The use of 'and so on' is an intuitive process which must be avoided within ZF. Having defined ω in this way, however, we can proceed to make deductions about the properties that its elements have. In Chapter 1 we listed five basic properties which we took as the common starting point for deductions about the set ℕ of natural numbers. These we called Peano's axioms. We can now demonstrate on the basis of the ZF axioms that our set ω indeed satisfies these axioms. This will mean that the elements of ω really do behave as natural numbers.

Theorem 4.12

$(\omega, {}^+, 0)$ is a model for Peano's axioms, i.e.

(i) (P1*) $0 \in \omega$.

(ii) (P2*) For each $x \in \omega$, $x^+ \in \omega$.

(iii) (P3*) If $x \in \omega$, then $x^+ \neq 0$.

(iv) (P4*) If $x, y \in \omega$ and $x^+ = y^+$, then $x = y$.

(v) (P5*) If A is a subset of ω which contains 0 and contains x^+ for every $x \in A$, then $A = \omega$.

Proof

(i) This is obvious.

(ii) We have already shown that $x \neq x^+$ for every set x.

(iii) For any set x, $x^+ = x \cup \{x\}$, so that $x \in x^+$. Hence $x^+ \neq \emptyset$, and so $x^+ \neq 0$.

(iv) Suppose that $x, y \in \omega$ with $x^+ = y^+$ and $x \neq y$. Then $x \cup \{x\} = y \cup \{y\}$. Now $x \in x \cup \{x\}$, so $x \in y \cup \{y\}$ also. Since $x \neq y$ we must have $x \in y$. Similarly $y \in x$. But this is impossible, by the foundation axiom (ZF9). See Exercise 18 on page 129.

(v) Let $A \subseteq \omega$ be such that $0 \in A$ and $x^+ \in A$ for every $x \in A$. Then A is a successor set. But ω is a subset of every successor set, so $\omega \subseteq A$. Hence, $A = \omega$.

Remarks

(a) The successor operation on ω is denoted by ${}^+$ rather than ′, which was the symbol used in Chapter 1, this is a difference merely of notation.

(b) The proofs of parts (ii) and (iv) above depended on axiom (ZF9). This dependence is inessential, as other proofs of these can be given using part (v) of the theorem and avoiding (ZF9). See Corollary 4.14 for one of these.

(c) The reader may be surprised and somewhat suspicious at the ease with which part (v), the principle of mathematical induction, is demonstrated above. There is no trick. The hard work has been done elsewhere, in the construction of the set ω. The principle of mathematical induction is a property of the set ω as we have defined it.

Notation

Now that we know that the elements of ω behave as natural numbers, we shall follow our use of bold face type for **0, 1, 2**, ... by using boldface letters **m, n, p, x, y**, etc. to denote abstract natural numbers, i.e. elements of ω. This will serve both as an intuitive aid when we deal with properties of abstract natural numbers and as a reminder that these are not in fact natural numbers in the ordinary sense, so we should be careful not to make unwarranted assumptions about them.

▶ In Chapter 1 we asserted that the operations of addition and multiplication could be defined using Peano's axioms, but we chose not to do that there. Instead, we gave as further basic properties of ℕ the inductive definitions of these operations. This rebounds on us here, for we are now required to verify that these are valid for our set ω of abstract natural numbers. We have to define these operations on ω and show that they have the necessary properties. This we shall do, but it requires application of a substantial theorem, which we shall give as Theorem 4.15. First, however, let us consolidate what we have so far, noting some properties of ω and its elements, and how the induction principle works in practice.

Theorem 4.13
(i)   For each $n \in \omega$, $n \in n^+$.
(ii) For each $n \in \omega$, $n \subseteq n^+$.
(iii) For each $n \in \omega$, $\bigcup n^+ = n$.

Proof
(i)   This has already been noted.
(ii) $n^+ = n \cup \{n\}$, and clearly $n \subseteq n \cup \{n\}$.
(iii) This is harder. It requires Theorem 4.12 (v).

Let $A = \{n \in \omega : \bigcup n^+ = n\}$. First of all, $0 \in A$, since $\bigcup 0^+ = \bigcup (0 \cup \{0\}) = \bigcup (\emptyset \cup \{\emptyset\}) = \bigcup \{\emptyset\} = \emptyset = 0$. Next, suppose that $n \in A$, i.e. that $n \in \omega$ and $\bigcup n^+ = n$. Then

$$\bigcup (n^+)^+ = \bigcup (n^+ \cup \{n^+\})$$

$$= (\bigcup n^+) \cup \bigcup \{n^+\} \quad \text{(think about it)}$$

$$= n \cup n^+$$

$$= n \cup (n \cup \{n\})$$

$$= n \cup \{n\}$$

$$= n^+.$$

Thus $n^+ \in A$ whenever $n \in A$. The set A therefore has the required properties, so we conclude that $A = \omega$, i.e. that the result holds for each $n \in \omega$.

Corollary 4.14
If $m, n \in \omega$ and $m^+ = n^+$, then $m = n$.

Proof

This result was given as Theorem 4.12 (iv), of course, but now we can prove it without recourse to axiom (ZF9). Suppose that $m, n \in \omega$ with $m^+ = n^+$. Then, using Theorem 4.13 (iii), we have $m = \bigcup m^+ = \bigcup n^+ = n$.

▶ Let us recall now the basic property given in Chapter 1 defining the addition operation on $\mathbb{N}$.

(A) There is a two-place function on $\mathbb{N}$ (denoted by +) with the properties:

$$m + 0 = m, \quad \text{for every } m \in \mathbb{N},$$

and

$$m + n' = (m + n)', \quad \text{for every } m, n \in \mathbb{N}.$$

We must verify this in respect of our set ω of abstract natural numbers and our successor operation. To do this we first validate a general procedure for constructing functions of which the above is a simple example, namely definition by induction. This is obviously related to the induction principle, but should not be confused with it. We shall need the *recursion theorem* (a consequence of the axioms of ZF) in order to guarantee the existence of functions defined by inductive schemes. Addition and multiplication will be particular cases to which we shall apply it. This theorem was referred to in Chapter 1, as it is related to Theorem 1.8. After reading the proof of Theorem 4.15, the reader should consider how to construct a proof of the more general Theorem 1.8.

Theorem 4.15 (recursion theorem)
Let X be any set, let $a \in X$ and let g be any function from X to X. Then there is a unique function f from ω to X (or a subset of X) such that

$$f(0) = a,$$

and

$$f(n^+) = g(f(n)) \quad \text{for each } n \in \omega.$$

Proof

The function f (if it exists) will be a subset of the Cartesian product $\omega \times X$. Let

$$u = \{z \in P(\omega \times X): (0, a) \in z \ \& \ (\forall y)(\forall t)((y, t) \in z \Rightarrow (y^+, g(t)) \in z)\}.$$

Clearly, $\omega \times X \in u$, so $u \neq \emptyset$. Let $f = \bigcap u$. Certainly, f is a subset of $\omega \times X$. We must show that it is a function and that its domain is ω. Let s be the domain of f. Then $s \subseteq \omega$, and $0 \in s$ since $(0, a) \in f$. Also, if $n \in s$ then $(n, t) \in f$ for some t, so $(n^+, g(t)) \in f$, and so $n^+ \in s$. Consequently (Theorem 4.12) $s = \omega$ as required, and it remains to show that f is single valued.

Let $r = \{y \in \omega: \text{there is precisely one } b \in X \text{ with } (y, b) \in f\}$. We know that $(0, a) \in f$. Suppose that $(0, c) \in f$, with $c \neq a$. Let $f' = f \backslash \{(0, c)\}$ (remove $(0, c)$ from f). Then $f' \in u$, since $(0, a) \in f'$ and whenever $(y, t) \in f'$ we have $(y^+, g(t)) \in f'$, because the discarded element cannot possibly be equal to $(y^+, g(t))$ for any y and t. But $f = \bigcap u$ and f' is a proper subset of f. However, $f' \in u$ implies that $\bigcap u$ is a subset of f'. Contradiction. Consequently $0 \in r$.

Next we must show that $y^+ \in r$ whenever $y \in r$. Suppose that there is precisely one $b \in X$ with $(y, b) \in f$. Then certainly $(y^+, g(b)) \in f$. Suppose also that $(y^+, d) \in f$ with $d \neq g(b)$. By an argument similar to the previous one we obtain a contradiction by considering $f'' = f \backslash \{(y^+, d)\}$. Details are left to the reader. By the principle of induction in ZF, then, we have $r = \omega$, and hence f is a function.

The inductive proof of uniqueness is left as an exercise.

▶ The development of arithmetic in ZF proceeds from the recursion theorem, because the introduction of addition and multiplication depends on it. We treat addition first.

Let $m \in \omega$ be fixed. We define a function s_m by an application of the recursion theorem in such a way for each $n \in \omega$

$$s_m(n) = m + n.$$

(s_m is the function '$m+$'.)

Definition

$s_m(0) = m,$

$s_m(n^+) = (s_m(n))^+,$ for each $n \in \omega.$

To justify this by the recursion theorem, all we need to verify is that $x \mapsto x^+$ is a function on ω. It is, for it is just the set $\{z \in \omega \times \omega : (\exists x)(\exists y) (z = (x, y) \& y = x^+)\}$, given by the separation axiom, and this set is easily shown to be a function.

We define *addition*, for arbitrary $m, n \in \omega$, by writing $m + n$ for $s_m(n)$. A moment's thought will convince the reader that this definition fits our intuition of what the sum of two natural numbers is. There is still much to be formally verified, however. For example, the commutative and associative laws for addition.

Theorem 4.16
(i) For each $n \in \omega$, $0 + n = n$.
(ii) For each $m, n \in \omega$, $m^+ + n = (m + n)^+$.
(iii) For each $m, n \in \omega$, $m + n = n + m$.
(iv) For each $m, n, p \in \omega$, $(m + n) + p = m + (n + p)$.
(See Theorem 1.3.)

Proof
(i) The proof is by induction on n, i.e. the formal proof uses the result given in Theorem 4.12, where the set concerned is $\{n \in \omega : 0 + n = n\}$. Details are left to the reader.

(ii) This is trickier. Let $A = \{y \in \omega : m^+ + y = (m + y)^+$ for every $m \in \omega\}$. Now $m^+ + 0 = s_{m^+}(0) = m^+ = (s_m(0))^+ = (m + 0)^+$, and so $0 \in A$. For the induction step, suppose that $y \in A$, i.e. that $m^+ + y = (m + y)^+$ for every $m \in \omega$. We have

$$m^+ + y^+ = s_{m^+}(y^+) = (s_{m^+}(y))^+ = (m^+ + y)^+$$

$$= ((m + y)^+)^+ = ((s_m(y))^+)^+ = (s_m(y^+))^+$$

$$= (m + y^+)^+.$$

Consequently, $y^+ \in A$. The induction principle (Theorem 4.12(v)) then ensures that $A = \omega$, as required.

(iii) Induction on m – left as an exercise.

(iv) Induction on p – left as an exercise.

▶ Multiplication is treated similarly:

Definition
For each $m \in \omega$ we can define a function p_m such that

$$p_m(0) = 0$$

$$p_m(n^+) = m + p_m(n), \quad \text{for each } n \in \omega.$$

We define *multiplication* by writing mn for $p_m(n)$ $(m, n \in \omega)$. Justification for the definition is given by the recursion theorem (Theorem 4.15) again, this time with the knowledge that $x \mapsto m + x$ (for fixed $m \in \omega$) is a function on ω.

▶ Properties such as the commutative, associative and distributive laws again have to be verified, but these are exercises in induction very much along the lines of Theorem 4.16, and we shall omit them.

Next we come to the ordering of the natural numbers, and just as in Chapter 1, we define $m < n$ to mean $(\exists x)(x \in \omega \ \& \ x \neq 0 \ \& \ m + x = n)$. This is another extension of our formal language. Properties of $<$ then follow from properties of addition – formal derivations merely reflect standard informal arguments.

Our abstract numbers, however, have some properties which we would not necessarily expect. These are properties which stem from the way that numbers have been defined as sets. Perhaps most significant is Theorem 4.18. First, a preliminary result.

Theorem 4.17
For every $m, n \in \omega$, $m \in m + n^+$.

Proof
We apply induction on n. Let $A = \{y \in \omega : m \in m + y^+ \text{ for all } m \in \omega\}$. For any $m \in \omega$, $m + 0^+ = (m + 0)^+ = m^+$ (by the definition of $+$), and so $m \in m + 0^+$, and hence $0 \in A$. Now suppose that $y \in A$, i.e. $m \in m + y^+$ for all $m \in \omega$. $m + (y^+)^+ = (m + y^+)^+$, and $(m + y^+)^+ = (m + y^+) \cup \{m + y^+\}$. Since $m \in m + y^+$, we have $m \in m + (y^+)^+$ also, as required, so that $y^+ \in A$. The result now follows, using Theorem 4.12.

Theorem 4.18
For any $m, n \in \omega$, $m < n$ if and only if $m \in n$.

Proof

Suppose that m, $n \in \omega$ and $m < n$, i.e. there exists $x \in \omega$ such that $x \neq 0$ and $m + x = n$. Now $x \neq 0$ implies that $x = p^+$ for some $p \in \omega$ (proof of this is left to the reader). Hence, $n = m + p^+$. By Theorem 4.17, $m \in m + p^+$, so $m \in n$, as required.

For the converse we use induction. Let A be the set $\{y \in \omega : (\forall z) ((z \in \omega \ \& \ z \in y) \Rightarrow z < y)\}$. First, note that $0 \in A$ trivially, since $z \in 0$ does not hold for any $z \in \omega$ (if the antecedent in an implication is false, then the implication as a whole is true). Next, suppose that $y \in A$, and consider an arbitrary $z \in \omega$ with $z \in y^+$. Now $y^+ = y \cup \{y\}$, so either $z \in y$ or $z = y$. In the former case we have $z < y$, so $z + u = y$ for some $u \in \omega$ with $u \neq 0$. Then $z + u^+ = y^+$ and, consequently, $z < y^+$. In the latter case, $z = y$, so $z + 1 = y^+$ and $z < y^+$ again, as required. This completes our inductive proof.

▶ Surprisingly, perhaps, we have shown that the relationship '$\in$' between numbers is the same as '$<$'. Also surprising is the following, which will be generalised later when we come to consider ordinal numbers.

Theorem 4.19

(i) For any m, n, $p \in \omega$, if $m \in n$ and $n \in p$, then $m \in p$.

(ii) For any m, $n \in \omega$, $m \in n$ implies $m \subseteq n$.

(iii) If m, $n \in \omega$, $m \neq n$ and $m \subseteq n$, then $m \in n$.

(iv) ω is a *transitive* set, i.e. if $n \in \omega$, then $n \subseteq \omega$.

(v) For each $n \in \omega$, $n = \{m \in \omega : m < n\}$.

Proof

(i) We can use the result of the last theorem. If $m + x = n$ and $n + y = p$ $(x, y \in \omega \backslash \{0\})$, then $m + (x + y) = p$, and so $m < p$, i.e. $m \in p$.

(ii) Let m, $n \in \omega$, with $m \in n$. Suppose that $k \in m$. By part (i), we have $k \in n$. Hence, $m \subseteq n$.

(iii) Left as an exercise (using induction on n).

(iv) This is also proved quite easily by induction on n.

(v) This follows from part (iv) together with Theorem 4.18.

▶ We have now almost reached the point in our formal development of numbers where we need proceed no further. The formal system has been developed in such a way that the standard mathematical notions and procedures have been formally described. Standard mathematical argument about these notions and procedures is in principle easily translated into the formal system, since mathematical reasoning consists of logical deduction. The constructions of the number systems serve as an illustration. The algebraic procedures of Chapter 1 could be carried out in ZF with little modification, since in ZF we now have available the notions of natural number, addition, multiplication, order, function, equivalence relation, equivalence class and, perhaps most used, the separation axiom.

There is one item previously mentioned (see Example 4.7) which we have not yet justified in ZF, and that is the construction of a set such as

$$\{x, P(x), P(P(x)), \ldots\}, \text{(where } x \text{ is any set)}$$

and of a function f with domain ω such that $f(n) = P^n(x)$. At first sight it would appear that the recursion theorem is sufficient, for we can write

$$\begin{cases} f(0) = x \\ f(n^+) = P(f(n)), \text{ for } n \in \omega. \end{cases}$$

The collection $\{x, P(x), P(P(x)), \ldots\}$ would then be a set, by the replacement axiom, being the image under the function f of the set ω.

This will not do however, because P is not a function. Given any set, $P(x)$ denotes its power set, but $x \mapsto P(x)$ cannot be a function because its domain would be the set of all sets, which we know cannot exist, by Exercise 9 on page 129. Certainly $x \mapsto P(x)$ (for $x \in X$) represents a function, if X is an arbitrary set, but that does not help us in this situation, since the domain would have to be (or at least contain) the collection $\{x, P(x), P(P(x)), \ldots\}$ which we are attempting to justify as a set. There is a vicious circle here. What we need is a general principle.

Theorem 4.20 (generalised recursion theorem)

Let $\mathscr{F}(x, y)$ be any formula of ZF such that for each set x there is precisely one set y for which $\mathscr{F}(x, y)$ holds. Then given any set a, there is a unique function f with domain ω such that

$$f(0) = a,$$

and

$$\mathscr{F}(f(n), f(n^+)) \quad \text{holds for each } n \in \omega.$$

Proof

First of all we show by induction that for each $n \in \omega$ there is a (uniquely determined) function f_n with domain $\{0, 1, \ldots, n\}$ such that

$f_n(0) = a$ and $\mathscr{F}(f_n(m), f_n(m^+))$ for each $m < n$. For the base step we can see that $\{(0, a)\}$ is the appropriate function. Now suppose that f_n is given.

$\mathscr{F}(x, y)$ determines a function, so there is a unique y such that $\mathscr{F}(f_n(n), y)$ holds. Obtain f_{n^+} from f_n by adjoining the ordered pair (n^+, y), i.e. let $f_{n^+}(n^+) = y$, and $f_{n^+}(m) = f_n(m)$ for all $m < n^+$. Then f_{n^+} has the required properties. That the f_n are uniquely determined is left as an exercise.

Now consider the formula $\mathscr{G}(u, y)$:

$$u \in \omega \ \& \ (\exists v)(v \text{ is a function with domain } u^+ \ \& \ v(0) = a \ \&$$

$$(\forall z)(z \in u \Rightarrow \mathscr{F}(v(z), v(z^+)) \ \& \ y = v(u)).$$

Translated, this says: $u = n$ for some $n \in \omega$ and $y = f_n(n)$. Because of the preceding discussion we can see that $\mathscr{G}(u, y)$ determines a function, and consequently, by the replacement axiom, the image of ω under it is a set, i.e., there is a set s whose elements are just all of the $f_n(n)$ for $n \in \omega$. Our required function f is then

$$\{(x, y) \in \omega \times s : x = n \ \& \ y = f_n(n)\},$$

given by the separation axiom. It may aid understanding to point out that the function f is in fact the union of all the functions f_n $(n \in \omega)$, since for each n, f_{n^+} extends f_n. (The union axiom could not be applied since we did not have a *set* it could be applied to.)

The inductive proof of uniqueness is left as an exercise.

Corollary 4.21
The collection $\{a, P(a), P(P(a)), \ldots\}$ is a set, for any given set a.

Proof
Let $\mathscr{F}(x, y)$ be the formula $y = P(x)$. This certainly determines a function in our formal sense. By Theorem 4.20 then, there is a function f such that the domain of f is ω, $f(0) = a$ and $f(n^+) = P(f(n))$ for each $n \in \omega$. The image of ω under this function is the required set. Indeed, the required set is found in the proof of the theorem as the set s.

Exercises
1. Let x, y be sets. Find $\bigcup(x, y)$ and $\bigcup\bigcup(x, y)$. Find also $\bigcup(x \times y)$.
2. Prove that $(a, b, c) = (x, y, z)$ implies $a = x$, $b = y$, $c = z$, for any sets a, b, c, x, y, z.

3. Prove from the definitions that, for any sets x, y, z, $x \times y \times z$ is the set of all ordered triples (a, b, c) with $a \in x$, $b \in y$, $c \in z$. Generalise.

4. Let R be a binary relation. Using (ZF6), show that the object $\{y : (\exists x)$ $((x, y) \in R)\}$ is a set. Explain why (ZF7) is not a consequence of this.

5. Derive a contradiction from the supposition that $\{(x, y) : x \sim y\}$ is a set.

6. Express by a well-formed formula of ZF the sentence: 'f is an injection'. Do the same for: 'f is a surjection' and 'f is a bijection'.

7. Justify composition of functions in ZF. In other words, given functions f and g, show that the object $g \circ f$ is a set and is a function.

8. Express by a well-formed formula of ZF: 'R is a total order relation'. Express similarly: 'R is a well-order'.

9. Let z be an equivalence relation on a set x. Justify the existence of the set of all equivalence classes (the quotient set).

10. Let x and y be sets. Justify the existence of the set of all bijections of the form $f : x' \to y'$ where $x' \subseteq x$ and $y' \subseteq y$.

11. Let x be a non-empty set. Deduce a contradiction from the supposition that $x \subseteq x \times x$. (Hint: consider the set $x \cup \bigcup x$, and use (ZF9).)

12. Prove, using Theorem 4.12(v), that $0 \in n^+$, for each $n \in \omega$.

13. Give inductive proofs for the following:
    (i) $0 + n = n$, for all $n \in \omega$.
    (ii) $m + n = n + m$, for all m, $n \in \omega$.
    (iii) $(m + n) + p = m + (n + p)$, for all m, n, $p \in \omega$.
    (iv) $mn = nm$, for all m, $n \in \omega$.
    (c) $(mn)p = m(np)$, for all m, n, $p \in \omega$.

14. Prove the following:
    (i) For all x, $y \in \omega$, $x + y = 0 \Rightarrow x = 0$ and $y = 0$.
    (ii) For all x, $y \in \omega$, $xy = 0 \Rightarrow x = 0$ or $y = 0$.

15. Give definitions by recursive schemes (similar to those for s_m and p_m) of exponentiation of natural numbers and of the factorial function.

16. Let X be a set, let $g : X \to X$ be a function, and let $a \in X$. Show that there is a function φ with domain ω such that $\varphi(n) = g^n(a)$. (Here g^n denotes the composition of g with itself n times, for $n > 0$, and $g^0(a) = a$.)

17. Let a be some fixed set.
    (i) Show that $\{a, a \times a, a \times a \times a, \ldots\}$ is a set, and that there is a function f with domain ω such that $f(n) = a^n$ ($= a \times \cdots \times a$ with n factors) for $n \in \omega$.
    (ii) Show that $\{a, \bigcup a, \bigcup\bigcup a, \ldots\}$ is a set and there is a function g with domain ω such that $g(n) = \bigcup^n a$, for $n \in \omega$.

18. Using the generalised recursion theorem, prove that $\{\omega, \omega^+, (\omega^+)^+, \ldots\}$ is a set.

4.4 Sets and classes

We noted earlier that the comprehension principle (given any property, there is a set of all objects having that property) leads to a contradiction, namely Russell's paradox. We found that $\{x : x \notin x\}$ cannot be a set. We also found (see the Exercises on page 129) that there are other 'collections' which cannot be sets, among them $\{y : y \sim x\}$ and $\{y : x \subseteq y\}$ (for any fixed set x), and the collection of *all* sets. These are therefore also illegitimate applications of the comprehension principle. The system ZF avoids the trouble by not admitting the comprehension principle, and so shunning such collections altogether. It is not possible within ZF to construct or even mention them.

There is, however, another standard way of formulating set theory, developed later than ZF, which does allow entities such as the above examples to be objects mentionable within the system. This system allows an (almost) unrestricted comprehension principle, and the objects of the theory are generally called *classes*. We shall see shortly how Russell's paradox can be avoided, and how sets are classes of a particular kind, so that the theory of sets will be subsumed within the theory of classes. The system dates from 1925 in its original version, but it is only comparatively recently that mathematicians in general have become aware of the possibility that collections may be too big to be sets and that some care has to be taken about such collections as $\{H : H$ is a group isomorphic to $G\}$, where G is a fixed group. It is one thing to note that such a collection is too big to be a set and is therefore a 'class'. It is another to be aware of what that entails: what are the properties of classes, and how can we tell whether a class is a set or a non-set?

The axiom system for class theory was developed by Bernays and by Gödel from an original version given by von Neumann in 1925. In the literature there are several different variants of it with different names (every possible permutation and combination of the initial letters of the names, it would seem). We shall describe one in which the contributions of von Neumann & Bernays are most significant, and call it VNB. This system has similarities with ZF (indeed, it was developed from ZF). Having examined ZF in some detail we shall not be as comprehensive in discussion of VNB, and we shall be slightly more formal in our exposition. The reader should understand the axioms, the differences between ZF and VNB, and the reasons for those differences. We shall also discuss briefly the usefulness of systems of class theory.

Before listing the axioms, let us investigate the way in which the comprehension principle can be allowed without leading to the contradiction in Russell's paradox. This is von Neumann's idea. Classes which are too large to be sets may be allowed (for example $\{x : x \notin x\}$), provided that these classes are not permitted to be *members* of other classes. So we can introduce a distinction between two sorts of classes: those which are elements of other classes, and those which are not. The first sort are called *sets*, and the second sort are called *proper classes*. The comprehension principle may now be expressed in the form: given any property that sets may or may not have, there is a *class* consisting of all *sets* which have the property. With this formulation we are safe from Russell's paradox, as we can now demonstrate.

Remark 4.22

Let A denote the class $\{x : x$ is a set and $x \notin x\}$. Now consider whether $A \in A$, as before. If $A \in A$ then A is a set and $A \notin A$, so we have a contradiction as before. If $A \notin A$ then A does not satisfy the conditions for belonging to A, i.e. it is not the case that both A is a set and $A \notin A$. By hypothesis, $A \notin A$, so the conclusion we must reach is that A is not a set. In our theory which admits non-sets (proper classes) there is no contradiction here. What happens is that A is a proper class and $A \notin A$.

▶ Now for the details of the system VNB. The basic undefined notions are *class*, *belongs to*, and *equals*. The objects of the theory are all classes – all elements of classes are classes, and all variables are presumed to stand for classes. For reasons which will become clear shortly, we shall use upper case letters X, Y, etc. as variables.

Definition

We define and introduce a special symbol $\mathcal{M}$ by letting $\mathcal{M}(X)$ stand for 'there is a class Y such that $X \in Y$'. As indicated above, we can think of $\mathcal{M}(X)$ as asserting 'X is a set'. (The use of the letter $\mathcal{M}$ arises from the German word for set, which is 'Menge'.)

▶ Although all objects are classes, we are defining a distinction amongst the classes between sets and non-sets. It is convenient to formalise this distinction by means of a suggestive notation. We use lower case letters x, y, etc. for variables when these are to stand for sets. In VNB a quantifier $(\forall X)$ says 'for all classes X'. We shall use the notation $(\forall x)$ to mean 'for all sets x'. More formally, if $\mathcal{A}(X)$ is some assertion about the unspecified class X,

$$(\forall x)\mathcal{A}(x) \text{ stands for } (\forall X)(\mathcal{M}(X) \Rightarrow \mathcal{A}(X)),$$

or, in words, for all classes X, if X is a set then $\mathcal{A}(X)$ holds. Similarly,

$$(\exists x)\mathcal{A}(x) \text{ stands for } (\exists X)(\mathcal{M}(X) \ \& \ \mathcal{A}(X)),$$

or, in words, there is a class X such that X is a set and $\mathscr{A}(X)$ holds. Also, we use lower case letters in formulas generally, using $\mathscr{A}(u)$ to abbreviate $\mathscr{A}(U)$ & $\mathscr{M}(U)$.

Axiom (VNB1) (extensionality axiom)

Two classes are equal if and only if they have the same elements.

This is the same as for ZF. Just as in ZF the notions involved lead to the introduction of the symbol $\subseteq$. We write

$$X \subseteq Y \text{ for } (\forall Z)(Z \in X \Rightarrow Z \in Y).$$

Axiom (VNB2) (null set axiom)

There is a *set* with no elements.

This is the same as for ZF, but notice that the axiom includes the assertion that the null set is indeed a set. Writing (VNB2) formally we would have

$$(\exists X)(\mathscr{M}(X) \,\&\, (\forall Y)(\neg(Y \in X))).$$

Here, as before, we introduce the symbol $\emptyset$ for the null set.

Axiom (VNB3) (pairing axiom)

Given any *sets* x and y, there is a *set* z whose elements are x and y.

Again this is identical with Axiom (ZF3), this being an assertion about sets in VNB. Notice that we cannot form unordered pairs of classes in general, since a proper class cannot be an element of any class. As before, $\{x, y\}$ denotes the unordered pair determined by x and y, $\{x\}$ abbreviates $\{x, x\}$, and (x, y) is defined to be $\{\{x\}, \{x, y\}\}$, where x and y are *sets*.

Axiom (VNB4) (union axiom)

Given any *set* x (whose elements are of course sets), there is a *set* which has as its elements all elements of elements of x.

Notice that this again is an assertion about sets in VNB, and that this axiom asserts more than existence. It asserts that the union of a set is a set.

The notations $\bigcup x$ and $x \cup y$ can be used as a consequence of this axiom (note the lower case letters), but, as with pairs, some care must be taken. As yet we have no axiom guaranteeing the existence of $\bigcup X$, where X is a proper class, though we shall include one later. Further, we cannot define $X \cup Y$ for classes X and Y via the unordered pair (recall that in ZF we let $x \cup y = \bigcup \{x, y\}$), since for proper classes unordered pairs do not exist. One of our later axioms will guarantee the legitimacy of $X \cup Y$.

Axiom (VNB5) (power set axiom)

Given any *set* x, there is a *set* which has as its elements all subsets of x.

This is about sets again, rather than classes in general. One of our later axioms will guarantee the existence of a *power class* for any given class, that is, the class of all its subsets. We use the same notation as before: $P(X)$ will denote the power class of X.

▶ Up till now the axioms of VNB have corresponded exactly with the axioms of ZF. The most substantial differences lie in the counterparts in VNB of (ZF6) and (ZF7). These will require a little discussion, so let us alter the order and give next the other axioms of VNB, those which correspond to (ZF8) and (ZF9).

Axiom (VNB6) (infinity axiom)

There is a *set* x such that $\emptyset \in x$, and such that for every set $u \in x$ we have $u \cup \{u\} \in x$.

Axiom (VNB7) (foundation axiom)

Every non-empty *class* X contains an element which is disjoint from X.

Note that this is an assertion about classes in general. (VNB7) serves exactly the same purpose in regard to classes as (ZF9) does for sets.

▶ Now we come to the interesting part. Recall that (ZF6) was redundant (it was a consequence of (ZF7)). In the context of VNB there is a corresponding assertion about subsets of given sets, but here there is no corresponding reason to include it amongst the axioms. We included (ZF6) in order to clarify the explanation then. There is no such reason here, so we move directly on to the counterpart of (ZF7).

Axiom (ZF7) is an axiom scheme, in effect infinitely many axioms, one for each formula $\mathscr{F}(x, y)$ which determines a function, and it states that given any set u there is a set consisting of all y such that $\mathscr{F}(x, y)$ holds for some $x \in u$. In VNB the replacement axiom will also state that the image of a set under a function is a set, but this can be expressed in a different way, in fact by a single axiom rather than an axiom scheme.

Axiom (VNB8) (replacement axiom)

Given any *set u* and any function F (i.e. class of ordered pairs which is single valued), there is a *set v* consisting of all sets y for which there is a set $x \in u$ with $(x, y) \in F$.

Notice that the function F may be a proper class. This enables the single axiom (VNB8) to cover all functions determined by formulas (in the sense of (ZF7)), as a consequence of (VNB9), which follows shortly. For a formula $\mathscr{F}(x, y)$ which determines a function, the collection

$$F = \{(x, y): x \text{ and } y \text{ are sets and } \mathscr{F}(x, y)\}$$

was not necessarily a set in ZF, but it may be allowed as a class in VNB. Thus a formula determining a function will actually correspond to a class F of ordered pairs which can be referred to within the system.

Examples 4.23

(a) $\{(x, y): x \text{ and } y \text{ are sets and } x = y\}$ is a proper class. Supposing it to be a set leads to a contradiction along the lines of Russell's paradox. It is a 'universal' identity function. This does not exist in ZF. Notice, however, that the formula $x = y$ in ZF is a formula which (in our terminology) determines a function.

(b) $\{(x, y): x \text{ and } y \text{ are sets and } y = \boldsymbol{P}(x)\}$ is likewise a proper class. The only way of mentioning this collection in ZF is by means of the formula $y = \boldsymbol{P}(x)$. (This is not a well-formed formula, but it can be seen to be equivalent to $(\forall z)(z \in y \Leftrightarrow (\forall u)(u \in z \Rightarrow u \in x))$.)

▶ We have left until last the most important axiom of VNB, where perhaps it should have come first. This was in order to emphasise the similarities between VNB and ZF, which are many. Several hints have been dropped as to the nature of (VNB9), so without further ado here it is.

Axiom (VNB9) (comprehension axiom)

Let $\mathscr{A}(X)$ be a well-formed formula in which the only quantifiers used are *set* quantifiers (as described above). Then there is a class consisting of all *sets* x for which $\mathscr{A}(x)$ holds.

Expressing this in perhaps a more familiar way: given a formula $\mathscr{A}(X)$ as described above, there is a class $\{x : x$ is a set & $\mathscr{A}(x)\}$. This of course is an axiom scheme, one axiom for each formula $\mathscr{A}(X)$.

▶ Notice the crucial difference between (VNB9) and (ZF6). The latter allowed subsets to be constructed from a given set, given various formulas. The former allows classes to be formed much more generally, given just the formulas. Of course (VNB9) includes (ZF6) in one sense: given a class Y and a formula $\mathscr{A}(X)$ as above, there is a *class* $\{x : x$ is a set & $x \in Y$ & $\mathscr{A}(x)\}$. So (VNB9) allows construction of *subclasses*. The force of (ZF6), that a formula determines a sub*set* of a given *set* is again here a consequence of the replacement axiom (see Theorem 4.9).

Axiom (VNB9) is not an unfettered comprehension principle. It is restricted in two ways. First, it allows the construction of the class of all *sets* with a given property. It is not allowed to construct the class of all *classes* with a given property. The reason for this restriction is because we are not allowing proper classes to be elements of other classes, in order to avoid Russell's paradox. Second, the formula $\mathscr{A}(X)$ is not allowed to contain class quantifiers. The reasons for this are more subtle. We mentioned the problem of impredicativity in relation to Axiom (ZF6) in Section 4.2. Similar considerations apply here. If $\mathscr{A}(X)$ contained a class quantifier then the class being determined would depend on properties of classes generally, including the one being determined. Next, it can be shown that the restriction to set quantifiers means that the axiom scheme (VNB9) is equivalent to finitely many of its instances, with the consequence that VNB, unlike ZF, can be specified by a finite number of axioms. This property of VNB has no great significance, but it can be of advantage to the logician who may wish to use it. And last, and possibly most significant, taking Axiom (VNB9) in this form yields the result of Theorem 4.25: in VNB we can prove exactly the same results *about sets* as we can in ZF. Without the restriction to set quantifiers, we would have a system of class theory in which would be provable more theorems about sets than in ZF. A detailed discussion of these matters may be found in the book by Fraenkel, Bar-Hillel & Lévy.

Let us now list some of the consequences of (VNB9).

Examples 4.24

(a) *Unions*

Given any class Z, by (VNB9) we can form the class

$$\{x : x \text{ is a set \& there is } y \in Z \text{ with } x \in y\}.$$

Here the formula $\mathcal{A}(x)$ is $(\exists y)(y \in Z \ \& \ x \in y)$. Notice that the quantifier is a set quantifier. (VNB9) thus guarantees the existence of the class $\bigcup Z$, where Z is any class. Remember, however, that (VNB4) is needed to guarantee that in the case where Z is a *set*, $\bigcup Z$ is also a *set*.

The union of *two* classes cannot be defined as for ZF, however. Recall that $x \cup y$ (for sets) was there defined as $\bigcup \{x, y\}$. Because unordered pairs exist only for sets, we cannot define $X \cup Y$ for classes in general by use of unordered pairs. Using (VNB9), however, given classes X, Y, we can write

$$X \cup Y = \{z : z \text{ is a set } \& \ (z \in X \vee z \in Y)\}.$$

(b) *Intersections*

Given any class Z, by (VNB9) we can form the class

$$\{x : x \text{ is a set } \& \ x \in y \text{ for every set } y \in Z\}.$$

This is denoted by $\bigcap Z$.

Also, given two classes X and Y, by (VNB9) we can form the class

$$\{z : z \text{ is a set } \& \ (z \in X \ \& \ z \in Y)\}.$$

This is denoted by $X \cap Y$.

(c) *Complements*

Given any class X, let

$$\bar{X} = \{y : y \text{ is a set } \& \ y \notin X\}.$$

Here is an essential difference between VNB and ZF. Absolute complements do not exist in ZF. For example, the complement of $\emptyset$ would be the collection of all sets. Notice that in VNB, there is a class $\bar{\emptyset}$, which is the class (usually denoted by V) of all sets. Certainly $\bar{\emptyset}$ must contain all sets, and it cannot contain anything else, for no proper class can belong to any class.

(d) *Power class*
Given any class X, let

$$P(X) = \{y : y \text{ is a set } \& \ y \subseteq X\}$$
$$= \{y : y \text{ is a set } \& \ (\forall z)(z \in y \Rightarrow z \in X)\}.$$

Observe that the power class $P(X)$ contains as elements all sub*sets* of X, not all subclasses, since proper classes cannot be elements of classes.

(e) *Cartesian product*
Given any classes X and Y, we can form the class

$$X \times Y = \{(u, v) : u \in X \ \& \ v \in Y\}.$$

This is not quite in the form required by (VNB9) so we must transform it. We require to find a well-formed formula $\mathscr{A}(z)$ so that $X \times Y = \{z : z \text{ is a set } \& \ \mathscr{A}(z) \text{ holds}\}$. First, we can write

$$X \times Y = \{z : (\exists u)(\exists v)(z = (u, v) \ \& \ u \in X \ \& \ v \in Y)\}.$$

Of course, it is implicit that in the above z must be a set (if u is a set and v is a set then (u, v) is a set). In the formula occurring above, we have now only to rewrite $z = (u, v)$ using only $\in$ and $=$. For those readers who like loose ends tied up, here is how it is done. By definition $(u, v) = \{\{u\}, \{u, v\}\}$, so $z = (u, v)$ is equivalent to

$$(\forall w)(w \in z \Leftrightarrow (w = \{u\} \text{ or } w = \{u, v\})).$$

Now $w = \{u\}$ is equivalent to

$$(\forall t)(t \in w \Leftrightarrow t = u),$$

and $w = \{u, v\}$ is equivalent to

$$(\forall t)(t \in w \Leftrightarrow (t = u \text{ or } t = v)).$$

Putting these together and inserting the result in place of $x = (u, v)$ in the above, yields the required formula $\mathscr{A}(z)$.

(f) *Membership relation*
There is a class consisting of all ordered pairs (x, y) for which x and y are sets and $x \in y$. To see this we apply a process similar to that for (e) above, to fit $\{(x, y) : x \in y\}$ into the form required for an application of (VNB9).

► Most of the above examples go to emphasise the analogy between VNB and ZF, at least with regard to normal set operations. VNB is a broader system than ZF, in the sense that the objects referred to in it are classes, which fall into two categories: sets and proper classes. Thus the two systems are not really equivalent, since results may be proved in VNB which not only cannot be proved in ZF – they may indeed have no counterpart in ZF. However, as regards assertions about *sets*, ZF and VNB are equivalent systems, in the following precise sense.

Theorem 4.25

Let $\mathscr{A}$ be a well-formed formula of ZF. Then $\mathscr{A}$ may be regarded as a formula of VNB if the variables are taken to be set variables and the quantifiers to be set quantifiers, and $\mathscr{A}$ is provable in ZF if and only if it is provable in VNB.

We cannot go into the proof, since it uses methods of mathematical logic which we have not covered. The implication one way: if $\mathscr{A}$ is provable in ZF then $\mathscr{A}$ is provable in VNB, is rather easier than the other, and the reader with some knowledge of logic may consider this part as an exercise.

► Thus ZF and VNB can serve equally well as systems of set theory – the same results about sets are derivable in both. There is one further observation to be made, however, about the relationship. Other results are derivable in VNB – can we be sure that none of these is contradictory? In other words: is VNB consistent? We cannot answer this question in absolute terms, for the reason which has been mentioned earlier, namely: on what principles could a proof of consistency depend? What we can say is the following.

Theorem 4.26

If ZF is consistent then VNB is consistent also.

Proof

There is a theorem of logic which states that in an inconsistent system, every statement is provable. This is a consequence of the fact that the propositional formula $(p\ \&\ \neg p)\Rightarrow q$ is a tautology, irrespective of what statements p and q stand for. Let us suppose that VNB is not consistent. Then there is a contradiction $p\ \&\ \neg p$ derivable from the axioms for VNB. Consequently there is a contradiction q derivable in VNB which involves only set variables and set quantifiers. By Theorem

4.25, this contradiction would be provable in ZF also, and so ZF is not consistent. Thus if VNB is not consistent, ZF is not consistent. The result we require now follows.

▶ The axiom of choice was mentioned in Section 4.2 as additional to the axioms of ZF. There is a corresponding axiom which may be added to those of VNB, and we state it here for the sake of completeness.

(AC) (axiom of choice, class form)

> Given any (non-empty) class X whose elements are non-empty disjoint sets there is a class which contains precisely one element from each set belonging to X.

▶ Let us end the section with some general comments about sets and classes. VNB successfully avoids Russell's paradox while allowing the use of the comprehension principle applied to sets. Thus it restores the intuitive process of collecting together which is one of the principal motivating ideas behind the notion of set. This probably seems more useful than it is. The contradictions inherent in the 'set of all sets' are avoided by calling it the 'class of all sets'. But what can we do with the class of all sets? The answer is: not very much, because it is not permitted to be a member of any class, and this means that no useful mathematical procedures can be carried out using it.

Another example may illustrate this point better. In VNB we can form the class of all sets which are equinumerous with a given set. This would appear to open the possibility of defining cardinal numbers as follows: given any set x, let card $x = \{y : y$ is a set and $y \sim x\}$. As we have seen, this makes sense, although card x is a proper class, for all non-empty sets x. In this way, each set corresponds to a unique cardinal number, and $x_1 \sim x_2$ if and only if card $x_1 = $ card x_2. But this is not a useful definition, for it would not be permissible to collect together cardinal numbers into classes. Requiring a mathematician to work with such objects is like tying an athlete's legs together. He just cannot work that way. He can work with objects only as long as he is allowed to discuss sets of those objects.

This example is given in order to illustrate the practical uselessness of classes as such. VNB is useful as a system of *set* theory, and it gives some respectability to the comprehension principle, but consideration of properties of classes as such is not useful. Indeed, there is a more general philosophical point to be made. Just as it can be argued that 'the collection of all sets' has some meaning, it can be argued that 'the collection of all proper classes' has some meaning. The former has been given some respectability in the system VNB. The latter clearly cannot be a set or a class. What then is it?

Exercises

1. What is $\{X\}$ if X is a proper class? What can (X, Y) mean in the different situations where X and Y are either sets or proper classes?
2. By finding appropriate well-formed formulas, use the comprehension axiom of VNB to justify the formation of the following classes:
   (i) $\{y : y$ is a set and $x \subseteq y\}$, where x is some fixed set.
   (ii) $\{(x, y) : x$ and y are sets and $\mathscr{F}(x, y)\}$, where $\mathscr{F}(x, y)$ is some well-formed formula of VNB in which the only quantifiers are set quantifiers.
   (iii) $\{y : y$ is a set and y is a binary relation$\}$.
   (iv) $\{f : f$ is a set and f is a function$\}$.
   (v) $\{y : x \sim y\}$, where x is some fixed set.
   Which of these are proper classes?
3. Prove that axioms (VNB8) and (VNB9) imply the following: given any well-formed formula $\mathscr{A}(y)$ in which the only quantifiers occurring are set quantifiers, and given any *set* x, there is a *set* $\{y \in x : y$ is a set and $\mathscr{A}(y)$ holds$\}$.
4. Prove that the intersection of a class and a set is a set. (Hint: use axiom (VNB8).) Deduce that every subclass of a set is a set. Can a class have a proper class as a subclass?
5. Prove that the Cartesian product of two sets is a set.
6. Find a well-formed formula of VNB which expresses the assertion 'X is an equivalence relation'. Justify the existence of equivalence classes, for any equivalence relation X. What can be said about the existence of a quotient class (i.e. the collection of all equivalence classes)?
7. Let $\mathscr{A}$ be a well-formed formula of ZF. Show that (regarding all the variables occurring in $\mathscr{A}$ as *set* variables) if $\mathscr{A}$ is provable in ZF then $\mathscr{A}$ is provable in VNB.

4.5 Models of set theory

We have deliberately played down the formal aspect of axiomatic set theory. But the point has been made that there is a distinction between provability from the axioms (of ZF, say) and actual (intuitively based) truth. The system of axioms is constructed with two aims: to enable as many truths as possible to be derived as consequences and to avoid having any contradictions as consequences. Now intuitively it is reasonable to believe that assertions about sets are either true or false (at least we may believe so about meaningful assertions), so there are two classes of assertions: those which are true and those which are not.

Any formal system of set theory which is proposed will divide the collection of all assertions about sets likewise into two: those which are provable and those which are not. The link between truth and provability in a formal system is one of the prime considerations of mathematical logic. This book is not about mathematical logic, so we shall not be concerned with detailed aspects of this link, but it is possible for us to examine broad issues, for they are very relevant to the purpose, usefulness and indeed limitations of formal set theory.

In 1930 Gödel proved his celebrated theorem about incompleteness of formal systems. We need not go so far as to state his theorem precisely. It will be sufficient to note its consequences for ZF (and similarly for VNB or any other axiomatic system of set theory which is adequate for the same purpose). The consequence is that there is necessarily an assertion about sets (a formula of ZF with no free variables) which is not derivable from the axioms of ZF and whose negation is not derivable either (provided, of course, that ZF is consistent). This spells trouble, for intuitively either this assertion is true or its negation is true, but neither is provable. There is thus an assertion about sets which is true but cannot be derived from the axioms of ZF. And the trouble is deeper than it may seem. This incompleteness applies to any consistent formal system of set theory (with a restriction that the set of axioms be recursive, but that need not concern us), so that we cannot escape the difficulty by including the true but underivable assertion as an additional axiom. There would be another true but underivable assertion in the augmented system.

There is, therefore, a fundamental limitation on formal axiomatic systems such as ZF. And 'incompleteness' is a good word to describe it. An axiom system for set theory can never provide the whole picture. Indeed, the idea of a common starting point for set theory takes a substantial knock at the hands of Gödel. It is one thing to write down and agree on some basic properties of sets (the axioms of ZF, for example). It is another, however, to expect that they will settle all questions relating to sets. As we have just seen, this expectation cannot be met.

There are perhaps two attitudes that we might take to this state of affairs. One is to try to develop the 'best possible' common starting point and to use it as such, namely, a useful codification of the intuitive properties of sets. The other (and these attitudes are by no means mutually exclusive) is to treat the axioms for set theory even more like axioms in abstract algebra. This is where the idea of models comes in.

Let us use the analogy of group theory. Take the assertion $(\forall x)(\forall y)$ $(xy = yx)$. This is not derivable from the axioms of group theory. We know this because there are groups in which it is false (non-abelian groups). Of course there are also groups in which it is true. Any assertion which is a logical consequence of the axioms for group theory is true in every group. Assertions which are not consequences of the group axioms may be true in some groups but not in others. So by analogy let us introduce the notion of a *universe* (of sets). A universe is a collection (whatever that means) of objects (which we may call *abstract sets*) satisfying the axioms of ZF. Then a group and a universe are both examples of *models* of formal axiomatic systems. Just as an underivable assertion (the commutative law) gives rise to groups with essentially different properties, the way is clear in principle to consider different universes with essentially different properties.

Let us try to put this in a different light. ZF is a formal system. The objects it refers to are thought of as sets because it was postulated originally as a common starting point for set theory. But in a logical system the symbols used have a purely formal existence separate from any *meaning* which may subsequently be attached to them. When we seek to find out more about ZF itself we need not regard it as a common starting point for mathematics – rather, we may regard it as a collection of axioms which characterise a type of abstract mathematical system. Indeed, we may regard a universe as a collection of objects of an unspecified nature but in which the axioms of ZF hold. This mirrors exactly the idea of an abstract group. The group theorist frequently is not concerned with what the *elements* of his groups are, as long as they have the properties laid down by the axioms.

There are certainly logical difficulties here, due to the apparent circularity, but with sufficient care they can be overcome. The essence of this is to work within the framework of a 'metatheory' of the theory of sets, since there must be a distinction between *sets*, of which our universes are examples, and *abstract sets*, which are the objects referred to in the formal theory and which are the elements of our universes. A metatheory is a body of assumptions, which may be formalised or may be intuitive, which enable discussion to take place and results to be derived *about* (rather than *within*) our formal system.

The usefulness of these ideas lies in questions of consistency and independence, and shortly we shall state some substantial results of great interest. But let us now see just how it is that different universes of abstract sets can arise. The case of the commutative law for groups is an example which generalises, and let us state theorems from mathematical logic which show how.

Theorem 4.27
An axiomatic system is consistent if and only if it has a model.

Theorem 4.28
If S is a consistent axiomatic system and $\mathscr{A}$ is a formula of S whose negation is not derivable from the axioms of S, then including $\mathscr{A}$ as an additional axiom yields a consistent system.

▶ The former is the more substantial result, particularly the statement that every consistent system has a model. This means that, corresponding to any collection of formal axioms, so long as they are consistent, there is a universe of sets in which all of the primitive notions of the axiom system are realised and in which all of the axioms hold. The latter theorem above just describes a certain way in which consistent systems may be formed.

The incompleteness theorem implies that there is a formula $\mathscr{A}$ of ZF such that neither $\mathscr{A}$ nor its negation is derivable in ZF (under the assumption that ZF is consistent). Under the same assumption, therefore, Theorem 4.28 provides us with two different consistent systems, one with $\mathscr{A}$ as an additional axiom and the other with $\neg\mathscr{A}$ as an additional axiom. Theorem 4.27 now yields two different models, i.e. universes of abstract sets. In one universe $\mathscr{A}$ holds and in the other universe $\neg\mathscr{A}$ holds.

The reader who is not familiar with these ideas may be becoming increasingly incredulous as we apparently move further away from reality. Sets are sets and there can be only one real universe of sets. So what can be the use of this artificial idea of different universes of abstract sets? Without denying that it is artificial, it is not hard to show how it is useful. We may take the axiom of choice to illustrate the point. Suffice to say at present that until recently there was uncertainty amongst mathematicians about whether it was true and acceptable. Gödel in 1938 gave a partial answer, namely, the following theorem.

Theorem 4.29
Given that ZF is consistent, the system obtained by including the axiom of choice as an additional axiom is also consistent.

▶ The consequence is that (AC) is acceptable in the sense that no contradiction can be deduced from it (and the ZF axioms), provided that no contradiction can be derived from the ZF axioms alone. More particularly, it is impossible to prove the negation of (AC) in ZF. Gödel's method of proof in essence is an application of Theorem 4.27. From a model of ZF he constructed a model of ZF in which (AC) holds. Consistency of ZF yields a model of ZF, from which a model of ZFC is constructed, and this yields the consistency of ZFC.

The other side of this coin was exposed in 1963. Given that ZFC is consistent, it may be asked whether (AC) is in fact a consequence of the ZF axioms. Ideas of models may be used to prove that it is not.

Theorem 4.30 (Cohen 1963)

Given that ZF is consistent, (AC) cannot be derived as a consequence of the ZF axioms.

▶ The essence of the proof is the construction, given the existence of at least one model of ZF (by Theorem 4.27), of a model of ZF in which (AC) does not hold. If (AC) were a consequence of the ZF axioms, then it would hold in every model because the axioms must hold in every model. Thus (AC) cannot be a consequence of the ZF axioms.

By these methods the logical position of the axiom of choice is made clear, in relation to the axioms of ZF. This is not necessarily helpful in deciding whether (AC) is a true statement, but it certainly clears the ground.

We give now some other examples of the sort of results that proofs using models can yield.

Theorem 4.31

(i) Given that ZF is consistent, the continuum hypothesis (see Chapter 2 page 71) is not a consequence of (AC) in ZF. Neither is (AC) a consequence of the continuum hypothesis.

(ii) In Chapter 3 we noted that the Boolean prime ideal theorem is a consequence of the axiom of choice. If ZF is consistent, then the axiom of choice is *not* a consequence in ZF of the Boolean prime ideal theorem.

▶ Models of ZF are thus seen to be useful tools in mathematical logic. But there is a distinctly unsatisfactory feeling about the idea that there can be universes of sets which are essentially different. Is not formal set theory supposed to reflect reality? If so, which of the possible universes of sets is the real one? There is no answer to this question, for it requires clarification as to what is meant by the real universe of sets. Formal set theory is of little help with this. It is a philosophical question. The notion of set is in practice very simple, and mathematics with a background of set theory is very convenient on an everyday level, but in principle the nature of sets is highly problematic, in the way that we have just seen, and there are inconveniences. One of these was mentioned in Chapter 1. Corollary 1.9 asserts that all models of Peano's axioms are isomorphic. It was pointed out then that this is a result *about* number theory, and has to be proved in the wider framework of a theory of sets. What such a proof tells us is only that, in any given universe of sets, all models of Peano's axioms are isomorphic. It leaves open the possibility that different universes of sets might contain models of Peano's axioms which are essentially different.

To conclude the chapter, let us return to the question: 'what is a set?'. We have established that there is a need for mathematicians to agree on the properties of sets in order for the notion to be useful. This leads to the listing of axioms representing a common starting point, with the intention of writing down sufficient basic properties to characterise completely what a set is (without actually anywhere stating what a set *is*, of course). The best that has been done in this line is represented by the systems ZF and VNB which we have discussed in detail. These systems are in practice very useful as characterisations of the notion of set. The axioms listed provide a common starting point which is adequate for the working mathematician. But neither system characterises the notion of set completely, in view of the fact that different models exist. Thus the question 'what are sets?' cannot be answered by 'objects which satisfy the ZF axioms'. Indeed, the same can be said not just of ZF and VNB, but of any other extended system which may be postulated to replace them.

Another aspect of the same thing is the following. Theorem 4.27 says that a consistent formal system has a model. A model is a set (we used the word 'collection' before, but there is no difference on an intuitive level between 'set' and 'collection'). Set theory purports to be about sets, but the set which is a model for the formal system of set theory cannot itself be an object which is referred to in the formal system, and certainly cannot itself be an element of the model. Thus whatever is meant by the word 'set', a model for formal set theory cannot contain all sets. Indeed, this is just another way of saying: there is no set of all sets.

Further reading

Cohen & Hersh [6] A readable account of the use of models of set theory.

Enderton [9] A straightforward book with a similar aims to those of this book, but with slightly different content.

Fraenkel, Bar-Hillel & Levy [10] A comprehensive treatment of axioms for set theory, containing much illuminating discussion. Quite technical.

Grattan–Guinness [11] This book contains an interesting article on the origins of set theory.

Halmos [12] This has been a standard work for twenty years, and is still useful and very readable.

Hamilton [13] An introduction to mathematical logic, with discussion of the place of logic in mathematics.

van Heijenoort [14] This volume contains original papers (translated) by many of the early contributors to set theory, and is of great interest.

Hersh [15] A common-sense view of the purpose and philosophy of mathematics.

5

THE AXIOM OF CHOICE

Summary
The axiom of choice is stated in several different forms, and examples are given of its application in familiar situations. Proofs are given of many results mentioned in Chapters 2 and 3 which require the axiom of choice. The equivalence of the axiom of choice with Zorn's lemma and with the well-ordering theorem is proved. Details are given of several applications of Zorn's lemma, and there is some discussion of the consequences of the well-ordering theorem. The last section deals with some of the less acceptable consequences of the axiom of choice and with some weak versions of it. Lists are given of equivalents of the axiom of choice and of some important consequences of it.

Chapters 2 and 3 are prerequisites for this chapter. Chapter 4 provides a useful formal context for the ideas of this chapter, but it is not essential.

5.1 **The axiom of choice and direct applications**
Every infinite set has an infinite countable subset. Let us imagine how a proof of this might proceed. Given an infinite set A, choose an element a_0 of A. Next, choose an element a_1 of A different from a_0. Next, choose $a_2 \in A \backslash \{a_0, a_1\}$, and so on. Since A is infinite, the process never ends. A sequence $a_0, a_1, a_2, \ldots$ of distinct elements of A is obtained, the elements of which constitute an infinite countable subset of A. Now this argument is certainly persuasive, and on an intuitive level it certainly justifies the conclusion. However, there is an informality about it which has disturbed mathematicians. The never ending process of successive choices cannot actually be carried out. Further, the principles on which the above proof is based are vague, to say the least. This informality and vagueness can be overcome (at some cost to intuitive

understanding) by appealing to the general principle which is the subject of this chapter. Before we state it, let us see how it works in practice.

Theorem 5.1
Every infinite set has an infinite countable subset.

Proof
Let A be an infinite set, and for each $n \in \mathbb{Z}^+$, denote by $S_n(A)$ the set of all sequences of length n of distinct elements of A. Choose one fixed sequence, say A_n, from each set $S_n(A)$. Then the set S consisting of all elements contained in the sequences A_n (for $n \in \mathbb{Z}^+$) is an infinite countable subset of A. It is clearly infinite since there is no bound on the lengths of the sequences A_n. That it is countable may be demonstrated from first principles (without appeal to Theorem 2.14) as follows. Denote the sequence A_n by $(a_{n1}, a_{n2}, \ldots, a_{nn})$ (for each $n \in \mathbb{Z}^+$). Define $f: \mathbb{N} \to S$ by letting f map the elements of the sequence $\frac{1}{2}n(n+1), \frac{1}{2}n(n+1) + 1, \ldots, \frac{1}{2}(n+1)(n+2) - 1$, respectively, to $a_{n1}, a_{n2}, \ldots, a_{nn}$, for each $n \in \mathbb{N}$. This function f is a surjection, and so S is countable by Corollary 2.8.

▶ In the above proof the infinite succession of choices is avoided. The use of 'and so on', which was essential to the preceding informal proof, and which we considered to be unsatisfactory, is eliminated. From each set $S_n(A)$ an element A_n is chosen. Not only is this reduced to one mathematical step; it is done in such a way that the choices are no longer dependent on one another.

The statement of the axiom of choice is as follows.

(AC) Given any (non-empty set) x whose elements are pairwise disjoint non-empty sets, there is a set which contains precisely one element from each set belonging to x.

The set whose existence is asserted is called a *choice set* for the set x.

A slightly different formulation, which is rather more easily applied, and which does not require the disjointness condition, is the following.

(AC') Given any (non-empty) set x whose elements are non-empty sets, there is a function f such that $f(a) \in a$ for each $a \in x$.

The function whose existence is asserted is called a *choice function* for the set.

In the proof of Theorem 5.1 the given set x was $\{S_n(A): n \in \mathbb{N}\}$, and the choice set was the set $\{A_n: n \in \mathbb{N}\}$. Alternatively, the choice function

f would be such that $f(S_n(A)) = A_n$. The set or function 'does the choosing'.

(AC) and (AC') are equivalent. Indeed, (AC') means exactly the same as (AC) when applied to a set of disjoint sets. However, (AC') is apparently stronger, so let us note how (AC') follows from (AC) (we leave as an exercise the demonstration that (AC') implies (AC)). Suppose that (AC) is true and that $\mathcal{X}$ is a non-empty set of non-empty sets (not necessarily disjoint). Let $\mathcal{Y} = \{\{X\} \times X : x \in \mathcal{X}\}$. Distinct elements of $\mathcal{Y}$ are disjoint. To see this, let $X_1, X_2 \in \mathcal{X}$ and let $(a, b) \in \{X_1\} \times X_1$ and $(a, b) \in \{X_2\} \times X_2$. Then $a \in \{X_1\}$ and $a \in \{X_2\}$, so that $a = X_1$, $a = X_2$. Consequently, $X_1 = X_2$. We can therefore apply (AC) to the set $\mathcal{Y}$. A choice set for $\mathcal{Y}$ consists of one element from each set $\{X\} \times X$, for $X \in \mathcal{X}$, and is therefore a set of ordered pairs of the form (X, x) where $x \in X$ (with $X \in \mathcal{X}$). Let f be a choice set for $\mathcal{Y}$. Then f is a function with domain $\mathcal{X}$ such that $f(X) \in X$ for each $X \in \mathcal{X}$, by the above; i.e. f is a choice function for $\mathcal{X}$.

▶ The axiom of choice in itself appears to express an intuitive truth, and most mathematicians today find it acceptable. There are perhaps three standpoints from which it may be doubted or criticised, however. The first is from consideration of its consequences, some of which are held by some mathematicians to be paradoxical, and we shall return to this later. The other two are on philosophical grounds. First, (AC') is an assertion about *every* set of non-empty sets, and we should be wary of claiming that it is obviously true unless we are absolutely sure of what 'every set of non-empty sets' means. The concept of 'set' itself is not one whose fundamental nature is universally agreed or understood. So it could be argued that (AC) is rather too sweeping to be even meaningful. Finally, (AC) can be criticised because of its *non-constructive* nature: however 'true' it may appear, it asserts the existence of a set without giving any indication either of how to construct that set or of what its elements are. It is sometimes argued that a set cannot be said to exist unless it is clear what its elements are or there is some method given by which membership of it can be tested. The axiom of choice is rejected by intuitionists because of this lack of constructiveness. It is not part of the purpose of this book to expound intuitionist theory (or to criticise it) and, consequently, we shall take this matter no further.

Whether (AC) is true or false, and whether it is acceptable or un-acceptable, are in one sense irrelevant to the mathematics of this book, for it is certainly of interest to mathematicians to find out the inter-relationships between principles in this area. It is of mathematical interest to know, for example, that the axiom of choice and Zorn's lemma are equivalent principles and that the Boolean prime ideal theorem (Theorem 3.22), though implied by (AC), is not equivalent to it. Further, it is of significance in relation to the intuitionist approach to mathematics to know whether or not individual theorems of mathematics depend on the axiom of choice. In essence, the remainder of this chapter is a brief outline of those results in the foundations of mathematics which do depend on the axiom of choice. Before proceeding further, however, let us be clear about the context in which we are working. A statement such as '(AC) is equivalent to Zorn's lemma' is not really meaningful unless it is clearly understood which other principles may be used in the demonstration of the equivalence. In this book we have avoided, as far as possible, formal deductions within axiom systems, and we shall continue to do so. Nevertheless, we can be quite specific and state that we assume a body of results about sets, namely, the common starting point consisting of the axioms of Zermelo–Fraenkel set theory which have been listed in Chapter 4. Thus the theorems of this chapter should perhaps all be prefaced by 'it is provable in ZF that', especially those which state that certain principles are consequences or equivalents of the axiom of choice. It is important to remember, of course, that (AC) was excluded from the list of axioms of ZF, so we nowhere assume (AC) unless explicitly stated.

We shall examine in turn the topics covered in Chapters 2 and 3 with regard to the application of the axiom of choice. First, we introduce a new notion, namely, the Cartesian product of an infinite family of sets. (The definition of a family was given in Chapter 4.)

Definition

The *Cartesian product* of a family F (with index set I) is the set of all functions f from I to the union of all the sets $F(i)$ with $i \in I$, such that $f(i) \in F(i)$ for all $i \in I$. It is denoted by ΠF.

This definition is designed to cover the situation of an infinite family, and it does not agree precisely with the more familiar notion of finite Cartesian product in terms of ordered pairs. However, let us see by means of an example how the two notions of Cartesian product are related.

Example 5.2

Let A and B be sets and let F be the family with index set $\{1, 2\}$ for which $F(1) = A$ and $F(2) = B$. We shall find a clear one–one correspondence between $A \times B$ and the Cartesian product ΠF. The elements of $A \times B$ are ordered pairs (a, b) with $a \in A$ and $b \in B$. The elements of ΠF are functions f from $\{1, 2\}$ to $A \cup B$ with $f(1) \in A$ and $f(2) \in B$. Thus we can associate each such f with the ordered pair $(f(1), f(2))$ of its values, an element of $A \times B$. Conversely, each pair $(a, b) \in A \times B$ corresponds to a function of the form specified, namely, g, where $g(1) = a$, $g(2) = b$.

Example 5.3

Let F be the family with index set $\mathbb{N}$ with $F(n) = \mathbb{R}$ for each n. The product ΠF is the set of all functions $f: \mathbb{N} \to \mathbb{R}$ (the union of all $F(n)$ is $\mathbb{R}$, and $f(n) \in F(n)$ is automatically satisfied). We may consider such a function represented by the sequence of its values:

$$f(1), f(2), f(3), \ldots,$$

and by analogy with Example 5.2 we may think of ΠF as $\mathbb{R} \times \mathbb{R} \times \mathbb{R} \times \cdots$. Perhaps more significantly we may think of ΠF as $\mathbb{R}^{\mathbb{N}}$ ($\mathbb{R}$ to the power of $\mathbb{N}$), which we in fact defined in Chapter 2 to mean the set of all functions from $\mathbb{N}$ to $\mathbb{R}$. These ideas therefore give some intuitive justification for that earlier notation. In general, we have the following.

Theorem 5.4

Let X be a set and let F be the family with index set I such that $F(i) = X$ for all $i \in I$. Then ΠF is the same set as X^I, i.e. the set of all functions from I to X.

▶ These ideas are necessary in order to state and apply another equivalent version of the axiom of choice.

(AC*) The Cartesian product of a non-empty family of non-empty sets is non-empty.

On the face of it this would appear to be obviously true, clearer even than (AC) and (AC'). Nevertheless, it is equivalent to the other two. A collection $\mathscr{X}$ of sets may be considered as a family F with index set $\mathscr{X}$, where $F(X) = X$ for each $X \in \mathscr{X}$. If the Cartesian product of such a family is not empty then any element of it is a choice function for the collection, so (AC*) implies (AC'). Conversely, given a family F of non-empty sets, indexed by I, let $\mathscr{X}_F$ be the collection of sets $\{F(i) : i \in I\}$. By (AC') there is a function f from $\mathscr{X}_F$ to the union of all these sets $F(i)$ such that $f(x) \in X$ (for $X \in \mathscr{X}_F$). Now define a choice function f^*

for the family F by

$$f^*(i) = f(F(i)).$$

Consequently, (AC′) implies (AC*).

(AC*) is rather surprising. This formulation makes it certain that the axiom of choice will play an essential part in the results that we obtain about cardinal numbers of products of sets. Without it we cannot even assume that Cartesian products of non-empty sets are non-empty. Of course, we may be able to obtain information in particular cases without reference to (AC*). For example, given a family F with index set I such that $F(i) = X$ for every $i \in I$, it is clear without (AC*) that $\Pi F \neq \emptyset$, provided that $X \neq \emptyset$. We can give an explicit description of an element of ΠF in this case. For if $x \in X$, then the constant function f, with $f(i) = x$ for every $i \in I$, is an element of ΠF.

▶ Next, let us consider another aspect of Chapter 2, namely, the definition of what is meant by an infinite set. In Chapter 2 a non-empty set was defined to be finite if there is a bijection between the set and $\{1, 2, \ldots, n\}$ for some positive integer n, and to be infinite otherwise. When we considered sizes of sets we observed that it is possible for a set to contain a proper subset which is equinumerous with the whole set. The example given (Example 2.2) was $\mathbb{N}$ with its subset $2\mathbb{N}$. This situation is clearly impossible with finite sets. So the question arises: is possession of a proper subset equinumerous with the whole set a necessary and sufficient condition for a set to be infinite? A set satisfying this condition is said to be *Dedekind infinite*. A Dedekind *finite* set is a set with no such subset.

Theorem 5.5
(i) If a set is Dedekind infinite then it is infinite.
(ii) The axiom of choice implies that every infinite set is Dedekind infinite.

Proof
(i) Let X be a Dedekind infinite set, and let A be a proper subset of X with a bijection $f: X \to A$. There exists $x \in X \setminus A$, for which necessarily $x \neq f^r(x)$ for all $r \geq 1$ (since $f^r(x) \in A$). Consequently, the elements $x, f(x), f(f(x)), \ldots$ are all distinct elements of X, for if $f^m(x) = f^n(x)$ (with $m \geq n$, say), then $f^{m-n}(x) = x$, since f is a bijection, and this contradicts our earlier assertion unless $m = n$. Therefore X cannot be finite.

(ii) First, we show that every infinite countable set is Dedekind infinite. Let A be infinite and countable and let $f : \mathbb{N} \to A$ be a bijection. Define a subset A_0 of A by $A_0 = \{f(2k) : k \in \mathbb{N}\}$. Then A_0 is infinite and countable and, consequently, equinumerous with A.

     Now let X be any infinite set. By Theorem 5.1, using the axiom of choice, X contains an infinite countable subset, Y, say, so write $X = Y \cup Z$ with $Y \cap Z = \emptyset$. By the above, Y contains a proper subset Y_0 which is equinumerous with Y, so let $g : Y_0 \to Y$ be a bijection. Then $Y_0 \cup Z$ is a proper subset of X which is equinumerous with X via the bijection h given by

$$h(x) = \begin{cases} g(x) & \text{if } x \in Y_0 \\ x & \text{if } x \in Z. \end{cases}$$

Corollary 5.6

A set is Dedekind infinite if and only if it contains an infinite countable subset.

Proof

The proof is the same as for Theorem 5.5, except that the first part of the proof of (ii) is not required and, consequently, (AC) is not required for this corollary. Compare this result with Theorem 5.1 which, of course, uses (AC) in its proof.

▶ We now mention a group of results, all consequences of (AC), following on the ideas of Chapter 2.

Theorem 5.7

The axiom of choice implies that, given any sets X and Y:
(i) Either there is an injection $X \to Y$ or there is an injection $Y \to X$, i.e. either $X \preccurlyeq Y$ or $Y \preccurlyeq X$.
(ii) There is a surjection $X \to Y$ if and only if there is an injection $Y \to X$.
(iii) Either there is a surjection $X \to Y$ or there is a surjection $Y \to X$.

Proof

(i) The most convenient proof of this uses Zorn's lemma, and it will be given in the next section (Theorem 5.13).

(ii) Let $f: X \to Y$ be a surjection. For each $y \in Y$ we denote by $f^{-1}(y)$ the set $\{x \in X: f(x) = y\}$, which is not empty since f is a surjection. By (AC') there is a function g from the set $\{f^{-1}(y): y \in Y\}$ to X such that $g(f^{-1}(y)) \in f^{-1}(y)$, for each $y \in Y$. Now define $h: Y \to X$ by $h(y) = g(f^{-1}(y))$. It can easily be verified that h is an injection.

Conversely, let $\phi: Y \to X$ be an injection, and let y_0 be any element of Y. Define $\psi: X \to Y$ by:

$$\psi(x) = \begin{cases} \phi^{-1}(x) & \text{if } x \text{ is in the image of } \phi \\ y_0 & \text{otherwise.} \end{cases}$$

Then ψ is certainly a surjection, as required. Note that the axiom of choice is not required for this latter part. (Compare this part of the proof with the proof of Corollary 2.8, and note that (AC) was not required there.)

(iii) This is an obvious consequence of (i) and (ii). Indeed, given (ii), it is clear that (i) and (iii) are equivalent. (Note, however, that (AC) is not required for the implication (i) $\Rightarrow$ (iii).)

Corollary 5.8 (law of trichotomy)
The axiom of choice implies that, given any sets X and Y, precisely one of the following holds.
  (i) X and Y are equinumerous.
  (ii) X strictly dominates Y.
  (iii) Y strictly dominates X.

Proof
This is an immediate consequence of Theorem 5.7 (i) together with the Schröder–Bernstein theorem (Theorem 2.18). (Recall that X strictly dominates Y if there is an injection from Y to X but no bijection between X and Y.)

▶ This result is one of the more acceptable consequences of the axiom of choice, in that it accords precisely with our intuition about sizes of sets. Indeed, historically it was accepted as self evident long before the axiom of choice was formulated. Note, however, the extreme generality and the non-constructiveness again. All we are given is the bare existence of either a bijection or an injection. As to how it may be specified or even which way it goes there is no information.

Exercises

1. Let X be an infinite set and let $\{x\}$ be any singleton set. Prove that $X \cup \{x\}$ is equinumerous with X. (Use Theorem 5.1.)

2. Show that (AC) implies the following:
   (i) $\aleph_0 \leqslant \kappa$, for every infinite cardinal number κ.
   (ii) $1 + \kappa = \kappa$, for every infinite cardinal number κ.

3. Prove (without using (AC)) that $2 + \kappa = \kappa$ implies $1 + \kappa = \kappa$, for any cardinal number κ.

4. Prove (by induction) that every finite collection of non-empty sets has a choice function.

5. Prove that (AC') implies (AC).

6. Prove (using (AC)) that if I is a (non-empty) index set, and for each $i \in I$, A_i is a non-empty set, where $A_i \cap A_j \neq 0$ for every pair of distinct elements i, j of I, then card $I \leqslant$ card A, where $A = \bigcup \{A_i : i \in I\}$.

7. Let A, B, C be non-empty sets, and let F be the family with index set $\{1, 2, 3\}$ such that $F(1) = A, F(2) = B, F(3) = C$. Describe a bijection between $A \times B \times C$ and ΠF.

8. Describe a bijection between $\mathbb{R}^n$ and the Cartesian product of the family F with index set $I = \{1, \ldots, n\}$, with $F(i) = \mathbb{R}$ for each $i \in I$.

9. Describe bijections between the following sets:
   (a) $\mathbb{N} \times \mathbb{N} \times \mathbb{N} \times \cdots$.
   (b) ΠF, where F is the family with index set $\mathbb{N}$ and $F(n) = \mathbb{N}$, for all $n \in \mathbb{N}$.
   (c) The set of all infinite sequences of elements of $\mathbb{N}$.
   (d) $\mathbb{N}^{\mathbb{N}}$.

10. Let F be a family of non-empty subsets of $\mathbb{N}$. Prove without using (AC) that $\Pi F \neq \emptyset$. Let G be any family of non-empty subsets of $\mathbb{Z}$. Prove without using (AC) that $\Pi G \neq \emptyset$.

11. Let F be a family of sets with index set I. Suppose that $\bigcap \{F(i) : i \in I\} \neq \emptyset$. Prove without using (AC) that $\Pi F \neq \emptyset$.

12. Let A, B be Dedekind finite sets. Prove that $A \cup B$ and $A \times B$ are Dedekind finite.

13. Let X, Y be sets with $X \leqslant Y$ and Y Dedekind finite. Show that X is Dedekind finite.

14. Prove, using (AC) that a given totally ordered set is well-ordered if and only if it contains no infinite descending chain.

15. Show without the axiom of choice that:
    (i) $1 + \aleph_0 = \aleph_0$.
    (ii) $1 + 2^{\aleph_0} = 2^{\aleph_0}$.
    (iii) $1 + 2^{2^{\aleph_0}} = 2^{2^{\aleph_0}}$.
    Generalise this result.

5.2 Zorn's lemma and the well-ordering theorem

There are two important principles which are each equivalent to (AC). These are:

Zorn's lemma
If X is any non-empty ordered set such that every chain in X has an upper bound in X, then X contains at least one maximal element.

Well-ordering theorem
Given any set X, there is a binary relation on X which well-orders X.

In this section we shall give a proof of the equivalence of these with (AC) (Theorem 5.17), but before doing that it will be useful to examine the application and usefulness of these equivalent principles. A greater familiarity with the techniques involved will aid understanding of the proof, which is quite difficult.

Zorn's lemma is an indispensable tool for mathematicians in most fields. We shall give examples of several applications in different areas. The well-ordering theorem as such is perhaps less relevant to the working mathematician. Its direct application lies in the foundations of mathematics, in the definition and use of cardinal and ordinal numbers of sets, for example. Notice that both Zorn's lemma and the well-ordering theorem are non-constructive in the same sense that (AC) is. They assert existence, in the one case of a maximal element in an ordered set and in the other case of a well-ordering relation, without giving any indication of how the object in question may be constructed, or even described.

Let us consider the well-ordering theorem first. In Chapter 3 we discussed different well-orderings of sets, and we noted on page 94 that well-orderings of uncountable sets are perhaps difficult to visualise. The case of the set $\mathbb{R}$ was used there as an example. The well-ordering theorem states that every set, no matter how large or how ill-defined, can be well-ordered. It follows that it is possible to list by means of a generalised counting procedure the elements of $\mathbb{R}$, or indeed of any other uncountable set. Notice that it is only the possibility that is asserted. We saw on page 94 why, in the case of $\mathbb{R}$, there can be no practical procedure for listing the elements.

Perhaps the most important applications of the well-ordering theorem concern ordinal and cardinal numbers. These will be dealt with in detail in the next chapter, but let us here give a little amplification. In Chapter 2 the notion of cardinal number was introduced in an informal way as a property that equinumerous sets share. Similarly, an *ordinal* number may be thought of as a property shared by order isomorphic well-ordered sets. Put another way, ordinal numbers will correspond with our generalised counting procedures. Two sets ordered by the same counting procedure (i.e. with elements paired off *in order*) will have the same ordinal number. The place of the well-ordering theorem in this is that it asserts that every set will be associated with some ordinal number through being well-ordered somehow. Of course, the situation is not exactly like that of cardinal numbers, for the same set may be well-orderable in many different ways, and so may correspond to many different ordinal numbers. Ordinal numbers play a very important part in the foundations of mathematics, and we shall give a full account of them (including discussion of cardinal numbers) in Chapter 6. For the moment, however, let us try to extend our intuitive ideas about well-ordered sets and counting procedures.

Example 5.9
As we saw in Chapter 3, the set $\mathbb{N}$ may be 'counted' in several different ways. For example, the following are listings by different generalised counting procedures, and each yields a well-ordering of $\mathbb{N}$ if we presume numbers to the left precede numbers to the right in the order.

(a) $0, 1, 2, 3, \ldots$.

(b) $1, 2, 3, 4, \ldots, 0$.

(c) $0, 2, 3, 4, \ldots, 1$.

(d) $2, 3, 4, 5, \ldots, 0, 1$.

(e) $1, 3, 5, \ldots, 0, 2, 4, 6, \ldots$.

(f) $1, 2, 4, 8, \ldots, 0, 3, 5, 6, 7, 9, \ldots$.

(g) $0, 1, 2, 4, 6, 8, \ldots, 3, 9, 15, 21, \ldots, 5, 25, 35, \ldots, 7, 49, \ldots, \ldots$

(the counting procedure here is: list first 0, 1 and all multiples of 2, then list all multiples of 3 not already occurring, then multiples of 5, and so on).

▶ Clearly, $\mathbb{N}$ can be listed in infinitely many different ways. Observe that (b) and (c) are isomorphic, as are (e) and (f). Listings isomorphic to (a) let us call standard single listings. Examples of these are 1, 0, 3, 2, 5, 4, ... and 2, 1, 0, 5, 4, 3, ... Each of the listings in Example 5.9 corresponds to an ordinal number. The ordinal number associated with a standard single list is usually denoted by ω. In the next chapter we shall see the connection between this ω and the symbol $\boldsymbol{\omega}$ used in Chapter 4.

A list such as (b) consisting of a standard single list with one further element has ordinal number $\omega + 1$. It makes a difference, clearly, which end the extra element is added on, however. An element added at the left-hand end of a standard single list yields an ordering isomorphic to (a) again. By analogy, (d) will have ordinal number $\omega + 2$, (e) and (f) will have ordinal number $\omega + \omega$, and (g) will have ordinal number $\omega + \omega + \omega + \cdots$, which may be denoted by $\omega \cdot \omega$ or ω^2.

Remark 5.10

For any finite set X, all possible well-orderings of X are isomorphic. It is intuitively clear that any two finite listings of the same length are isomorphic, because the elements can be matched up in order. (See Exercise 1 on page 183.) It is only when a set is infinite that possibilities arise for different generalised counting procedures. Thus finite well-ordered sets are isomorphic if and only if they are equinumerous, so finite ordinal numbers may be taken to be identical with finite cardinal numbers.

▶ The well-ordering of sets also has significance through the method of proof by transfinite induction. This depends on the following result: given any non-empty set X, well-ordered by the relation R, if for $x \in X$, x has property P whenever all predecessors of x (under R) have property P, then every element of X has property P. The proof of this is not difficult – it depends only on the definition of a well-order – and the reader may wish to attempt a proof before we return to this topic later in the book. See Theorem 6.15.

Lastly, a well-ordering of a set enables us to pick out particular elements, since least elements always exist, and this is where the link with the axiom of choice comes. Given a collection of non-empty sets, assuming that the union of all of them can be well-ordered, we can define a choice function f by taking $f(X)$ to be the least element in X, for each set X in the collection. Thus (AC') is a consequence of the well-ordering theorem.

Next, let us consider Zorn's lemma. This principle provides a sufficient condition for the existence of a maximal element in a given ordered set. Its most common application is in cases where the ordered set in question is a collection of sets ordered by $\subseteq$

A chain in such an ordered set is a subset which is totally ordered by $\subseteq$. It may happen that for each such subset, its union is a member of the original set, and if so it is clearly an upper bound for the subset. This is frequently a convenient way of verifying the condition required in Zorn's lemma.

Theorem 5.11

Zorn's lemma implies that every vector space contains a basis.

Proof

First, note that a basis is a maximal linearly independent subset, i.e. a linearly independent set which is not contained in any larger linearly independent set. So consider the collection $\mathscr{X}$ of all linearly independent subsets of a given vector space V. $\mathscr{X}$ is ordered by $\subseteq$. Let $\mathscr{C}$ be a chain in $\mathscr{X}$. We shall show that the union of the sets in $\mathscr{C}$ is a member of $\mathscr{X}$, i.e. is linearly independent, and hence that $\mathscr{X}$ contains an upper bound for $\mathscr{C}$. Let $v_1, v_2, \ldots, v_n \in \bigcup \mathscr{C}$, and let $a_1, a_2, \ldots, a_n$ be scalars such that

$$a_1 v_1 + \cdots + a_n v_n = 0.$$

There must exist sets $C_1, \ldots, C_n \in \mathscr{C}$ with $v_1 \in C_1, \ldots, v_n \in C_n$. Here is where we use the fact that $\mathscr{C}$ is a chain, for the set $\{C_1, \ldots, C_n\}$ must have a greatest element (under $\subseteq$), say C_k, and it follows that $v_1 \in C_k$, $v_2 \in C_k, \ldots, v_n \in C_k$. Now C_k is linearly independent, so we must have $a_1 = a_2 = \cdots = a_n = 0$. Thus $\bigcup \mathscr{C}$ is linearly independent, and the hypotheses of Zorn's lemma are satisfied. Hence, there exists in V a maximal linearly independent subset, i.e. a basis.

Theorem 5.12 (Boolean prime ideal theorem)

Zorn's lemma implies that every Boolean algebra contains a prime (equivalently, maximal) ideal. (This result was previously stated without proof, as Theorem 3.22.)

Proof

Let A, ordered by R, be a Boolean algebra, and let $\mathscr{X}$ be the set of all proper ideals in A, ordered by $\subseteq$. Suppose that $\mathscr{C}$ is a chain in $\mathscr{X}$. We show that $\bigcup \mathscr{C}$ is an ideal of A. First note that $\bigcup \mathscr{C}$ is not empty, since each ideal of A is necessarily non-empty. Next, let $a \in \bigcup \mathscr{C}, b \in \bigcup \mathscr{C}$. Then $a \in C_1$ and $b \in C_2$, say, where $C_1 \in \mathscr{C}$ and $C_2 \in \mathscr{C}$. Since $\mathscr{C}$ is a chain, we must have either $C_1 \subseteq C_2$ or $C_2 \subseteq C_1$. Suppose the former, without loss of generality, so that $a \in C_2$ and $b \in C_2$. Then $a \vee b \in C_2$, since C_2 is an ideal. Consequently, $a \vee b \in \bigcup \mathscr{C}$. Last, suppose that $b \in \bigcup \mathscr{C}$ and aRb. Then $b \in C$, say, with $C \in \mathscr{C}$, and since C is an ideal, we have $a \in C$. Hence, $a \in \bigcup \mathscr{C}$, as required. We have now verified that $\bigcup \mathscr{C}$ is an ideal of A. Also $\bigcup \mathscr{C}$ is a proper ideal, since $1 \notin \bigcup \mathscr{C}$. If $1 \in \bigcup \mathscr{C}$, we would have $1 \in C$ for some $C \in \mathscr{C}$, but the elements of $\mathscr{C}$ are proper ideals, so this is impossible. Hence, $\bigcup \mathscr{C}$ is an upper bound for $\mathscr{C}$ in $\mathscr{X}$. Zorn's lemma now yields the existence of the required maximal ideal.

▶ Similar direct applications of Zorn's lemma may be used elsewhere in algebra to demonstrate the existence of maximal objects. Some examples are given in the exercises at the end of this section. However, it is not always clear on the face of things that Zorn's lemma will be applicable. The following theorem, answering a question we have considered already, is an example of such a situation (see Theorem 5.7).

Theorem 5.13

The axiom of choice implies that, given any two non-empty sets X and Y, either $X \leqslant Y$ or $Y \leqslant X$.

Proof

Let $\mathscr{X}$ be the set of all bijections $f : A \to B$ with $A \subseteq X$ and $B \subseteq Y$. A bijection, of course, is a function, and a function is a set of ordered pairs, so we can regard $\mathscr{X}$ as a collection of sets, ordered by $\subseteq$. We shall verify the hypotheses of Zorn's lemma with respect to $\mathscr{X}$, so let us first see how the existence of a maximal element in $\mathscr{X}$ gives us the desired conclusion. Let $f_0 : A_0 \to B_0$ be maximal in $\mathscr{X}$. Then either $A_0 = X$ or $B_0 = Y$, for otherwise both $X \backslash A_0$ and $Y \backslash B_0$ are non-empty, and there exist $a \in X \backslash A_0$ and $b \in Y \backslash B_0$. Therefore f_0 may be extended by adjoining the ordered pair (a, b) to obtain a bijection from $A_0 \cup \{a\}$ to $B_0 \cup \{b\}$, contradicting the maximality of f_0. We have shown, then, that either $A_0 = X$, in which case f_0 is an injection from X to Y, and we have $X \leqslant Y$, or $B_0 = Y$, in which case f_0^{-1} is an injection from Y to X, and we have $Y \leqslant X$.

It remains, therefore, to verify the hypotheses of Zorn's lemma in this situation. The sets X and Y are non-empty, and it follows that $\mathscr{X}$ is non-empty since there is certainly a bijection between $\{x_0\}$ and $\{y_0\}$, for any choice of $x_0 \in X$ and $y_0 \in Y$. As we have seen, $\mathscr{X}$ is ordered by $\subseteq$. Let $\mathscr{C}$ be a chain in $\mathscr{X}$. We show that $\bigcup \mathscr{C}$ belongs to $\mathscr{X}$ and is therefore the required upper bound. Certainly $\bigcup \mathscr{C}$ is a set of ordered pairs (x, y) with $x \in X$ and $y \in Y$, so it is a binary relation from X to Y. To see that $\bigcup \mathscr{C}$ is a function, suppose that $(x, y) \in \bigcup \mathscr{C}$ and $(x, z) \in \bigcup \mathscr{C}$. Then we have $(x, y) \in f_1$ and $(x, z) \in f_2$, say, with $f_1 \in \mathscr{C}$ and $f_2 \in \mathscr{C}$. Since $\mathscr{C}$ is a chain we may presume without loss of generality that $f_1 \subseteq f_2$. It follows that $(x, y) \in f_2$ and $(x, z) \in f_2$, and since f_2 is a function we must have $y = z$. Thus $\bigcup \mathscr{C}$ is a function, from a subset of X to a subset of Y. Moreover, $\bigcup \mathscr{C}$ is an injection. To show this, let $(x_1, y) \in \bigcup \mathscr{C}$ and $(x_2, y) \in \bigcup \mathscr{C}$ and show by an argument exactly analogous to the preceding one that $x_1 = x_2$. This is left as an exercise. Lastly, we can regard

$\bigcup \mathscr{C}$ as a bijection onto the set of elements of Y actually occurring as function values. This completes the demonstration that $\bigcup \mathscr{C} \in \mathscr{X}$ and the theorem is therefore proved.

▶ Theorem 5.7(i) is therefore proved, as promised, and its consequences, mentioned earlier, are now justified.

Theorem 5.13 can be expressed in terms of cardinal numbers, according to the practice of Chapter 2.

Corollary 5.14

Given any two sets X and Y, either card $X \leqslant$ card Y or card $Y \leqslant$ card X.

▶ There is one other result mentioned in Chapter 2 which we can now prove by application of Zorn's lemma. The ideas of the proof will not be used subsequently, so it can be passed over without prejudice to later material.

Theorem 5.15

The axiom of choice implies that, for any infinite set A, $A \times A$ is equinumerous with A.

Proof

Let A be an infinite set. Then by Theorem 5.1, A has a countable infinite subset, C, say. We know (Theorem 2.12) that $C \sim C \times C$, so the set $\mathscr{X}$ of all bijections $B \to B \times B$, where B is a subset of A, is not empty. Regard $\mathscr{X}$ as an ordered set, with order relation $\subseteq$. Let $\mathscr{C}$ be a chain in $\mathscr{X}$. Just as in the proof of Theorem 5.13, we can show that $\bigcup \mathscr{C}$ is a function and is an injection. This part is left as an exercise. The domain of $\bigcup \mathscr{C}$ is a subset of A, say X. We show that $\bigcup \mathscr{C}$ maps X onto $X \times X$. Let $a \in X, b \in X$. Then there exist $f_i, f_j \in \mathscr{C}$ such that $a \in$ domain f_i and $b \in$ domain f_j. Since $\mathscr{C}$ is a chain we can say without loss of generality that $f_i \subseteq f_j$, so a and b both lie in domain f_j. But $f_j : X_j \to X_j \times X_j$ is a bijection, for some $X_j \subseteq A$, so $(a, b) \in X_j \times X_j$, and consequently $(a, b) = f_j(c)$ for some $c \in X_j$. Therefore (a, b) is the image of c under the function $\bigcup \mathscr{C}$ also, and we have shown that $\bigcup \mathscr{C}$ maps X onto $X \times X$. The hypotheses of Zorn's lemma are therefore verified, and the conclusion is that there exists a maximal element in $\mathscr{X}$. Denote it by $f : M \to M \times M$, say. We prove that $M \sim A$, so that $A \sim M \sim M \times M \sim A \times A$, and the required result follows. Suppose that it is not the case that $A \sim M$. We derive a contradiction. First we show that $M \preccurlyeq A \backslash M$. Otherwise, by Theorem 5.13, we would have $A \backslash M \preccurlyeq M$, and applying the lemma below, we would obtain $(M \cup A \backslash M) \sim M$, i.e. $A \sim M$, contrary to our supposition.

Lemma

For any sets P and Q, if $P \times P \sim P$ and $Q \preccurlyeq P$, then $P \cup Q \sim P$. (See Exercise 5 on page 72.)

Now, since $M \preccurlyeq A \backslash M$ there is a subset of $A \backslash M$ equinumerous with M, say N. We show that $M \cup N \sim (M \cup N) \times (M \cup N)$ via an extension of f, contradicting the maximality of f. Observe that $(M \cup N) \times (M \cup N) = (M \times M) \cup (M \times N) \cup (N \times M) \cup (N \times N)$, since M and N are disjoint. But

$$M \times N \sim N \times N \sim N \quad \text{(since } M \sim N),$$

$$N \times M \sim N \times N \sim N,$$

and

$$N \times N \sim N,$$

so by repeated use of the lemma, we have

$$(M \times N) \cup (N \times M) \cup N \sim N,$$

and hence

$$(M \times N) \cup (N \times M) \cup (N \times N) \sim N.$$

(Notice that $Q \sim P$ implies $Q \preccurlyeq P$.)

We now have bijections

$$f : M \to M \times M$$

and (say)

$$g : N \to (M \times N) \cup (N \times M) \cup (N \times N).$$

Since M and N are disjoint, and $M \times M$ is disjoint from $(M \times N) \cup (N \times M) \cup (N \times N)$, adjoining f and g yields a bijection from $M \cup N$ to $(M \cup N) \times (M \cup N)$, as required to give us our contradiction.

Corollary 5.16

Let κ be the cardinal number of any infinite set. Then the axiom of choice implies that $\kappa \kappa = \kappa$. The results of Theorem 2.38 are therefore verified as consequences of (AC).

▶ The remainder of this section is devoted to a proof of the equivalence of (AC), Zorn's lemma and the well-ordering theorem. The proof is lengthy and technical and may be omitted without prejudice to what comes later. A shorter proof will be given in Chapter 6, using methods which are not available to us yet. (See Example 6.22 and Exercise 7 on page 220.)

Theorem 5.17
The following principles are equivalent.
  (i) (AC).
 (ii) Zorn's lemma.
(iii) The well-ordering theorem.

The proof proceeds in two stages. First, we show that Zorn's lemma is equivalent to an apparently weaker principle, and then we prove that this principle is equivalent to (AC) and to the well-ordering theorem.

Theorem 5.18
Zorn's lemma is equivalent to the following modified principle: if X is any non-empty ordered set such that every chain in X has a least upper bound in X, then X contains at least one maximal element.

Proof
This modified principle is clearly implied by Zorn's lemma. We therefore need demonstrate the other implication only. Let X be a set, ordered by a relation R, such that every chain in X has an upper bound in X. Now let $\mathscr{X}$ be the set of all chains in X. Certainly $\mathscr{X}$ is ordered by inclusion. Let $\mathscr{C}$ be a chain in $\mathscr{X}$. We show that $\bigcup \mathscr{C} \in \mathscr{X}$, i.e. $\bigcup \mathscr{C}$ is a chain in X. Certainly $\bigcup \mathscr{C}$ is a subset of X. Let $a, b \in \bigcup \mathscr{C}$. Then $a \in C_1$ and $b \in C_2$ for some $C_1 \in \mathscr{C}$, $C_2 \in \mathscr{C}$. Since $\mathscr{C}$ is a chain we have $C_1 \subseteq C_2$ or $C_2 \subseteq C_1$, and we may suppose the former without loss of generality. Then $a, b \in C_2$, and C_2 is a chain in X, so we have aRb or bRa, as required to show that $\bigcup \mathscr{C}$ is a chain in X. Thus the chain $\mathscr{C}$ has an upper bound in $\mathscr{X}$, but note that $\bigcup \mathscr{C}$ is not merely *an* upper bound, it must be the least upper bound – $\bigcup \mathscr{C}$ is the smallest set containing as subsets all elements of $\mathscr{C}$. We can therefore apply the modified principle to $\mathscr{X}$. Let C_0 be a maximal element of $\mathscr{X}$. Then C_0 is a chain in X, so has an upper bound in X, say u. Then u is a maximal element of X, for if there existed $x \in X$ with uRx and $u \neq x$, then $C_0 \cup \{x\}$ would be a chain in X which strictly contains C_0, contradicting maximality of C_0. This completes the proof of Theorem 5.18.

► For the next stage, we prove three implications.
  (a) The well-ordering theorem implies (AC).
  (b) Zorn's lemma implies the well-ordering theorem.
  (c) (AC) implies the modified version of Zorn's lemma given in Theorem 5.18.

Proof of (a)

This has already been outlined. Given any collection of sets we specify a choice function f by means of the well-ordering theorem as follows. First form the union of all sets in the collection. This union can be well-ordered. For each set X in the collection, let $f(X)$ be the least element of X in this well-order.

Proof of (b)

Given any set X, we must show that there exists an order relation on X which well-orders X. Let W be the collection of all well-ordered subsets of X, i.e. all pairs (A, R) where $A \subseteq X$ and R is a well-order on A. We shall apply Zorn's lemma to W, but of course we must have an order on W for this to be possible or useful. This order is more complicated than in previous examples. We say that $(A, R) \leqslant (B, S)$ if the latter is an upward extension of the former, i.e. if $A \subseteq B$, R is the restriction of S to A, and S is such that every element of A precedes every element of $B \backslash A$. This may be expressed by saying that A is an initial segment of B. It can easily be shown (exercise) that $\leqslant$ is an order relation on W. To apply Zorn's lemma we must show that every chain in W has an upper bound in W, so let $\mathscr{C}$ be a chain in W. The union of the sets in $\mathscr{C}$ will provide our upper bound, but we require it to be appropriately well-ordered. It is in effect automatically well-ordered, for if a and b belong to the union then there exist (C_1, R_1) and (C_2, R_2) in $\mathscr{C}$ with $a \in C_1$ and $b \in C_2$. Without loss of generality we may suppose that $(C_1, R_1) \leqslant (C_2, R_2)$, so $C_1 \subseteq C_2$, and C_2 contains both a and b. Thus a and b are related under R_2.

We say that aRb if aR_2b. That this procedure is well-defined, yielding an ordering R of the union of the sets in $\mathscr{C}$, that R is a well-ordering, and that this union, ordered by R, is an upper bound for $\mathscr{C}$ must all be verified, but we shall omit the detail. Zorn's lemma gives the existence of a maximal element (M, R_0), say, in W. It must happen that $M = X$, for if not we may choose $x \in X \backslash M$, and well-order $M \cup \{x\}$ by the relation R_0', where $R_0' = R_0 \cup \{(a, x) : a \in M \cup \{x\}\}$. The effect of this is that every element of M precedes x in the ordering R_0'. Then $(M \cup \{x\}, R_0')$ belongs to W, and extends (M, R_0), contradicting the maximality of (M, R_0). Thus $X = M$, and R_0 is a well-ordering of X, as required.

Proof of (c)

Let X be a non-empty set, ordered by the relation R, such that every chain in X has a least upper bound in X. We require to show that there is a maximal element in X. For each element $x \in X$ let $S(x) = \{y \in X : xRy$ and $x \neq y\}$.

Notice that $S(x)$ may be empty – indeed, $S(x) = \emptyset$ if and only if x is maximal in X. Suppose that there is *no* maximal element in X and therefore that $S(x) \neq \emptyset$ for all $x \in X$. We shall derive a contradiction. (AC') may be applied to the collection $\{S(x) : x \in X\}$ of non-empty sets, so let F be a choice function for this collection. Thus $F(S(x)) \in S(x)$ for each $x \in X$. Now define a function $f : X \to X$ by $f(x) = F(S(x))$ $(x \in X)$. Notice that f has the property that $xRf(x)$ and $x \neq f(x)$, for each $x \in X$. Using our hypotheses about X we shall show that there is $z \in X$ such that $z = f(z)$, to give us our contradiction. From here on, a will be some fixed element of X. Intuitively, what we do is construct a set which contains $a, f(a), f^2(a), \ldots$, and which contains the least upper bound, a_1 (say), for this chain, and $f(a_1), f^2(a_1), \ldots$, and the least upper bound for this chain, etc. This process generates a chain which contains its own least upper bound, say a_0, and contains $f(a_0)$ also. But then necessarily $f(a_0)Ra_0$. It follows that $f(a_0) = a_0$, since we know that $a_0Rf(a_0)$ by the definition of f.

Now let us make the above argument precise. Let $\mathcal{X}_a$ be the collection of all subsets B of X satisfying

(i) $a \in B$.

(ii) $x \in B$ implies aRx.

(iii) $x \in B$ implies $f(x) \in B$.

(iv) for each subset C of B which is a chain in X the least upper bound of C is a member of B also.

Recall our intuitively generated set above and note that it will satisfy all of these requirements. We avoid the intuitive construction by considering rather the intersection of all the sets in $\mathcal{X}_a$, i.e. the smallest set satisfying all these conditions. For this to make sense we must be sure that $\mathcal{X}_a$ is not empty, so we observe that the set $\{x \in X : aRx\}$ satisfies conditions (i) to (iv). Thus $\bigcap \mathcal{X}_a$ makes sense, and we denote this set by A. It is a straightforward exercise to verify that $A \in \mathcal{X}_a$ (i.e. A satisfies (i) to (iv)). We shall show that A is a chain in X. When that is done, we may deduce that A has a least upper bound, say a_0, in X, and, by condition (iv), that $a_0 \in A$. By condition (iii), then, $f(a_0) \in A$, and consequently $f(a_0)Ra_0$. But certainly $a_0Rf(a_0)$, by the definition of f, so $a_0 = f(a_0)$ since R is an order relation, and this is the contradiction which will complete our proof.

It remains to fill in the gap by proving that A is a chain in X. This is rather lengthy and complicated. First, let

$$A^* = \{x \in A : y \in A \ \& \ yRx \ \& \ y \neq x \text{ imply } f(y)Rx\}.$$

Thinking of A as our intuitively generated set leads us to believe that $A^* = A$, since it would appear then that for no y is there any element lying strictly between y and $f(y)$. We shall show that $A^* = A$. By the same token, for each $b \in A^*$ the set

$$A_b = \{x \in A : xRb \text{ or } f(b)Rx\}$$

intuitively should be equal to A. To show this we verify that A_b satisfies (i) to (iv). For condition (i) note that $b \in A$, so aRb, since A itself satisfies condition (ii), and so $a \in A_b$. Condition (ii) is immediate. For condition (iii), let $x \in A_b$. Then xRb or $f(b)Rx$. Also $xRf(x)$. If $x = b$ then $f(b)Rf(x)$. If xRb and $x \neq b$ then since $b \in A^*$ we have $f(x)Rb$. Last, if $f(b)Rx$ then $f(b)Rf(x)$ by transitivity of R. In any case, therefore, we have $f(x)Rb$ or $f(b)Rf(x)$, i.e. $f(x) \in A_b$. For condition (iv), let C be a chain in A_b. Then C is a chain in A, so the least upper bound, say c_0, for C lies in A, since A satisfies condition (iv). $C \subseteq A_b$, so either xRb for all $x \in C$, or there exists $c \in C$ with $f(b)Rc$. In the former case we have c_0Rb necessarily, and in the latter case $f(b)Rc_0$, by transitivity. Consequently, $c_0 \in A_b$ as required. Thus A_b satisfies conditions (i) to (iv), and since $A_b \subseteq A$ it follows that $A_b = A$, by the original definition of A.

Now we show that $A^* = A$ by a similar procedure. For condition (i), note that aRx for every $x \in A$, so yRa and $y \neq a$ cannot hold for any $y \in A$, and $a \in A^*$ is vacuously true. As before, condition (ii) is automatically satisfied. For condition (iii), let $x \in A^*$. Then, as we have shown above, $A_x = A$, so for each $y \in A$, either yRx or $f(x)Ry$. Now let $y \in A$, $yRf(x)$, $y \neq f(x)$. Then we cannot have $f(x)Ry$, so we must have yRx. If $y = x$ then $f(y) = f(x)$ and $f(y)Rf(x)$ trivially. If $y \neq x$ then since $x \in A^*$ we have $f(y)Rx$ and then $f(y)Rf(x)$ follows by transitivity of R. Hence, $f(x) \in A^*$, as required. Last, for condition (iv), let C be a chain in A^*. Then C is a chain in A and its least upper bound, say c_0, lies in A. We show that $c_0 \in A^*$. Let $y \in A$, yRc_0, $y \neq c_0$. Now $A_b = A$ for each $b \in A^*$, so $A_c = A$ for each $c \in C$. Hence, $y \in A_c$ for every $c \in C$, and so either yRc or $f(c)Ry$, for every $c \in C$. Thus either $f(c)Ry$ for every $c \in C$ or there exists some $c_1 \in C$ with yRc_1. In the former case cRy for every $c \in C$, so c_0Ry, which is contrary to the original supposition about y. In the latter case, if $c_1 \neq y$ then $f(y)Rc_1$ (since $c_1 \in A^*$) and so $f(y)Rc_0$ since c_1Rc_0. And finally, if $c_1 = y$ then since $c_0 \neq y$ we have $c_1 \neq c_0$ and there exists $c_2 \in C$ with yRc_2, $c_2 \neq y$ (and c_2Rc_0 of course). As before, then, $f(y)Rc_2$, so $f(y)Rc_0$. We have proved that $c_0 \in A^*$ and the demonstration that A^* satisfies (i) to (iv) is complete. It follows that $A^* = A$ just as in the case of A_b.

To complete the proof that A is a chain in X, let $x, y \in A$. Then $x \in A^*$ and $y \in A_x$, since $A = A^* = A_x$. This means that either yRx or $f(x)Ry$. Since $xRf(x)$ we obtain the desired conclusion that either yRx or xRy.

The proof of Theorem 5.17 is now complete.

Exercises

1. Prove by induction that for any $n \in \mathbb{N}$, all well-ordered sets with n elements are isomorphic.

2. Show that any countable set can be well-ordered (without assuming (AC) or any equivalent principle).

3. Prove the principle of transfinite induction: if X is a non-empty set, well-ordered by the relation R, and A is a subset of X such that for each $x \in X$ we have $x \in A$ whenever all predecessors of x lie in A (i.e. $\{y \in X : yRx \text{ and } y \neq x\} \subseteq A$ implies $x \in A$), then $A = X$.

4. Prove (without assuming (AC) or any equivalent principle) that every non-empty set of non-empty well-ordered sets has a choice function. (Note that the sets must have given well-orderings – it is not sufficient for the sets to be well-orderable, for in that case well-ordering relations would have to be chosen, requiring (AC).)

5. Prove that Zorn's lemma implies the following:
   (i) In a commutative ring with unity every proper ideal is contained in a maximal proper ideal.
   (ii) Every lattice which contains a greatest element and at least one other element contains a maximal ideal (Theorem 3.24).
   (iii) Given any Boolean algebra A and any subset S of A, which does not contain the least element of A, there exists a maximal ideal in A which is disjoint from S.
   (iv) Given any set X and any binary relation R on X, there is a maximal (under $\subseteq$) subset Y of X such that $Y \times Y \subseteq R$.
   (v) Given any set X and any order relation R on X, there is a total order relation S on X such that $R \subseteq S$.

6. Show that the following principles are all equivalent to Zorn's lemma.
   (i) Every non-empty ordered set contains a maximal chain (maximal under $\subseteq$). (This result is known as Kuratowski's lemma.)
   (ii) If X is a non-empty set ordered by the relation R, in which every chain has an upper bound, and if a is any element of X, then X contains a maximal element x_0 such that aRx_0.

7. Prove, using (AC) or an equivalent principle, that there is a subset B of $\mathbb{R}$ such that
   (a) B is rationally independent, i.e. if $b_1, \ldots, b_n \in B$ and $q_1, \ldots, q_n \in \mathbb{Q}$, with $q_1 b_1 + q_2 b_2 + \cdots + q_n b_n = 0$, then $q_i = 0$ for $1 \leqslant i \leqslant n$.
   (b) B spans $\mathbb{R}$, i.e. given any $x \in \mathbb{R}$ there exist $b_1, \ldots, b_k \in B$ and $q_1, \ldots, q_k \in \mathbb{Q}$ such that $x = q_1 b_1 + \cdots + q_k b_k$.
   (Such a set B is called a *Hamel basis* for $\mathbb{R}$. Although (AC) guarantees the existence of a Hamel basis, it gives no help in constructing or describing one, and indeed none is explicitly known.)
8. Fill in the gaps left in the proof of Theorem 5.15.
9. Using Theorem 2.38, prove the following (assuming (AC)), where $\kappa, \lambda, \mu, \nu$ are cardinal numbers.
   (i) $\kappa < \lambda$ and $\mu < \nu$ imply $\kappa + \mu < \lambda + \nu$.
   (ii) $\kappa < \lambda$ and $\mu < \nu$ imply $\kappa\mu < \lambda\nu$.
   (iii) $\kappa + \mu < \lambda + \mu$ implies $\kappa < \lambda$.
   (iv) $\kappa\mu < \lambda\mu$ implies $\kappa < \lambda$.
10. Prove (assuming (AC)) that, for every infinite cardinal number κ, $\kappa + \aleph_0 = \kappa$ and $\kappa \aleph_0 = \kappa$.
11. In the proof of Theorem 5.17, show that the set A (defined as $\bigcap \mathscr{X}_a$) satisfies conditions (i) to (iv), and is therefore an element of $\mathscr{X}_a$.
12. Let X be a set totally ordered by the relation R. Recall (Exercise 20 on page 96) that a subset Y of X is cofinal in X if for each $x \in X$ there exists $y \in Y$ with xRy. Prove, using Zorn's lemma, that every totally ordered set contains a cofinal well-ordered subset (well-ordered by the inherited ordering).
13. (Harder) Prove that the following principle is equivalent to Zorn's lemma:
    If X is a non-empty ordered set such that every well-ordered subset has an upper bound in X, then X contains a maximal element. (Hint: use the principle given in Exercise 6(i) above.)
14. Assuming the axiom of choice, prove that for any infinite set X, the set of all finite subsets of X is cardinally equivalent to X.
15. Prove that a Hamel basis must have cardinal number $\aleph$.

5.3 **Other consequences of the axiom of choice**

In this section we shall outline some of the wide-ranging consequences of (AC) and discuss their logical relationships and their acceptability. In particular we shall consider some restricted forms of (AC) and their applicability. Most proofs will be omitted, and there are two reasons for this. The first is that we can obtain an overall view better that way, and the second is that many of the results presented here require methods from other branches of mathematics and from logic and axiomatic set theory which are beyond the scope of this book. Indeed, some of the results of this section have been obtained only very recently. The interested reader is referred to the book by Jech for a comprehensive exposition.

We start with two of the surprising consequences of (AC).

Theorem 5.19
The axiom of choice implies that there exists a set of real numbers which is not Lebesgue measurable.

Proof
Recall that Lebesgue measure μ has the following properties: the measure of a closed interval is just its length, μ is invariant under translations, i.e. $\mu(X) = \mu(\{x + a : x \in X\})$ for any $a \in \mathbb{R}$, and μ is countably additive, i.e. the measure of the union of a countable collection of disjoint sets is the sum of the measures of the sets.

Let us define a relation $\backsimeq$ on the interval $[0, 1]$ in $\mathbb{R}$ by: $x \backsimeq y$ if and only if $x - y$ is a rational number. It is an easy exercise to verify that $\backsimeq$ is an equivalence relation. Apply (AC) to the collection of equivalence classes to obtain a set T containing one number from each equivalence class. For each rational number q let

$$T_q = \{x + q : x \in T\},$$

i.e. T_q is the set obtained by translating T a distance q. Now the sets T_q are pairwise disjoint, and their union is all of $\mathbb{R}$. To see this, first suppose that $T_q \cap T_r \neq \emptyset$, so that

$$x + q = y + r \text{ where } x, y \in T \text{ and } q, r \in \mathbb{Q}.$$

This implies that $x - y = r - q \in \mathbb{Q}$, so $x \backsimeq y$. But distinct elements of T come from different equivalence classes, so we must have $x = y$ and, consequently, $q = r$, so $T_q = T_r$. Second, let $a \in \mathbb{R}$. Then $a = x + m$ for some $x \in [0, 1]$ and $m \in \mathbb{Z}$. Moreover, there must exist $y \in T$ such that $x \backsimeq y$, and hence $x = y + q$ for some $q \in \mathbb{Q}$. Consequently, $a = y + q + m$, so $a \in T_{q+m}$.

The set T is not Lebesgue measurable. Let us suppose the contrary. Now $\mathbb{R} = \bigcup_{q \in \mathbb{Q}} T_q$, a countable disjoint union. So $\mu(\mathbb{R}) = \sum_{q \in \mathbb{Q}} \mu(T_q)$. But $\mu(T_q) = \mu(T)$ for each $q \in \mathbb{Q}$. It follows that $\mu(T) > 0$, for if $\mu(T) = 0$ we would have $\mu(\mathbb{R}) = 0$, which is absurd. But $\mu(T) > 0$ leads to a contradiction also, as follows. For $q \in \mathbb{Q}$ with $0 \leqslant q \leqslant 1$ we have $T_q \subseteq [0, 2]$, and so $\bigcup \{T_q : q \in \mathbb{Q} \text{ and } 0 \leqslant q \leqslant 1\} \subseteq [0, 2]$. But the set on the left is an infinite countable union of sets with the same measure $\mu(T) > 0$, so it has infinite measure. The set on the right is an interval so its measure is its length, i.e. 2. A set with infinite measure cannot be contained in a set with finite measure, so our result is now demonstrated.

▶ The existence of a non-measurable set can be (and widely is) regarded as just one of the awkward properties of the set of real numbers. It does not run counter to intuition. However, the next result really does seem paradoxical.

Theorem 5.20 (the Banach–Tarski paradox)

The axiom of choice implies that a closed three-dimensional solid ball may be split into finitely many pieces which can be rearranged without distortion to form two solid balls of the same size as the original one.

We omit the proof of this. It includes an argument similar to the one in the previous proof but applying to rotations of the ball rather than translations of the line. The axiom of choice is used to select representatives from equivalence classes again. Of course, the proof, using (AC), is non-constructive and thus gives no indication of how the decomposition may be carried out, and certainly gives little geometrical insight into the form of the pieces. A proof may be found in the book by Jech.

▶ As has been observed, (AC) is a very strong assertion, since it refers to *any* collection of non-empty sets. Some of the results we have derived from (AC) may be shown to require only a weak version. Generally, a proof using (AC) will apply it to a particular set of sets, so what is applied is (AC) for that set of sets. For example, in the proof of Theorem 5.1, we used (AC) to select one sequence A_n from each of the sets $S_n(A)$ of n-element sequences, and clearly the following restricted version of (AC) would have sufficed:

(CAC) Every non-empty countable set of non-empty sets has a choice function.

(This principle is called the *countable axiom of choice*.)

There are other circumstances when (CAC) is strong enough to derive particular results. Certainly the cases of Zorn's lemma and the well-ordering theorem are not such, since these are equivalent to the full axiom of choice. As another example, let us return to Theorem 2.14 and give a proof without the sleight of hand referred to originally.

Theorem 5.21

The countable axiom of choice implies that the union of a countable set of countable sets is countable.

Proof

Let $A_0, A_1, A_2, \ldots$ be countable sets. A set is countable if and only if there is a surjection from $\mathbb{N}$ onto the set. For each $i \in \mathbb{N}$ let S_i denote the set of all surjections from $\mathbb{N}$ to A_i. By (CAC) there exists a set $\{f_i : i \in \mathbb{N}\}$ of functions such that $f_i \in S_i$ for each $i \in \mathbb{N}$. Now we can define a surjection $f : \mathbb{N} \to \bigcup_{i \in \mathbb{N}} A_i$ as follows. Given $n \in \mathbb{N}$, if $n > 0$ we may write $n = 2^k \times 3^l \times m$, where m is not divisible by either 2 or 3, and k and l are then uniquely determined by n. So let

$$\begin{cases} f(0) = f_0(0) \\ f(n) = f_k(l) \text{ if } n = 2^k \times 3^l \times m \text{ as above.} \end{cases}$$

Theorem 5.22

The countable axiom of choice implies that every infinite set is Dedekind infinite.

Proof

An easy exercise (see Theorem 5.5).

Example 5.23

In mathematical analysis, properties such as continuity and compactness are often defined in two different ways, either through what is commonly known as the ε–δ procedure or through limits of sequences. The equivalence of such forms of definition is often taken as immediate, whereas in fact the countable axiom of choice is involved. To illustrate this, consider the idea of limit point (on the real line, for the sake of simplicity). Let A be a subset of $\mathbb{R}$. The element $x \in \mathbb{R}$ is a *limit point* of A if

(i) there is a sequence $a_1, a_2, \ldots$ of elements of A which converges to x,

or

(ii) every neighbourhood of x contains an element of A distinct from x.

(A point satisfying (ii) is sometimes called an *accumulation point*.) (CAC) is required to show that the second definition implies the first. The converse is trivial, without the need for any form of the axiom of choice. Verification is left as an exercise.

► The idea behind the countable axiom of choice may be generalised in an obvious way. For each cardinal number κ we have the statement:

(AC$_\kappa$) If $\mathcal{X}$ is a non-empty set of non-empty sets, and if card $\mathcal{X} \leqslant \kappa$, then $\mathcal{X}$ has a choice function.

The countable axiom of choice is then the assertion (AC$_{\aleph_0}$). Each (AC$_\kappa$) on its own is strictly weaker than (AC) itself, and clearly, in any particular application of (AC), it will be a particular (AC$_\kappa$) which is applied. Again, of course, this remark does not apply to a general application of (AC) such as the proof of Zorn's lemma, where the full axiom of choice is required. Examination of the proof of Zorn's lemma from (AC) yields the following result, however.

Theorem 5.24

For any cardinal number κ, (AC$_{2^\kappa}$) implies that if $\mathcal{X}$ is an ordered set with card $\mathcal{X} = \kappa$ and such that every chain in $\mathcal{X}$ has an upper bound in $\mathcal{X}$, then $\mathcal{X}$ contains a maximal element.

A similar result is the following.

Theorem 5.25

For any cardinal number κ, (AC$_\lambda$) implies that every set with cardinal number κ can be well-ordered, where $\lambda = 2^{2^{\kappa^2}}$.

▶ Another weak version of the axiom of choice is the *principle of dependent choices*:

(DC) Let R be a binary relation on a non-empty set A, with the property that for each $x \in A$, the set $\{y \in A : xRy\}$ is not empty. Then there exists a sequence $x_0, x_1, x_2, \ldots$ of elements of A such that $x_n R x_{n+1}$ for each $n \in \mathbb{N}$.

(Note that there is no assumption about the relation R being reflexive, transitive, symmetric or anti-symmetric.)

(DC) allows a sequence of choices to be made, where each element chosen is 'dependent on' (i.e. related to) the previous one. It is not immediately apparent that (AC) implies (DC) but we summarise the situation (without proofs) in the next theorem. For an application of (DC) see Exercise 6 on page 191.

Theorem 5.26

(i) (AC) implies (DC).
(ii) (DC) implies (CAC).
(iii) Neither of the above implications can be reversed.

▶ Because of the controversy which has surrounded the axiom of choice, mathematicians have search for other principles which would help to decide questions in this area, perhaps through being more constructive, perhaps through being less general. There is one which perhaps deserves mention, since it has come into prominence recently. This is the so-called *axiom of determinateness*. This is couched in terms of games, and we shall state it thus, but we shall omit all detail of its application, so the ideas behind the statement of the axiom are not important for us. Given any set S of infinite sequences of natural numbers, the game G_S is described as follows. There are two 'players', and each in turn chooses a natural number, thus generating a sequence $n_0, n_1, n_2, \ldots$ (player 1 chooses $n_0, n_2, \ldots$, and player 2 chooses $n_1, n_3, \ldots$). If the resulting infinite sequence belongs to S then player 1 wins; otherwise player 2 wins. A strategy is a function from finite sequences of natural numbers to $\mathbb{N}$, and a player plays according to a strategy if he chooses at each stage the number given by this function applied to the finite sequence of numbers previously chosen. A strategy is a winning strategy if the player employing it wins, regardless of the actions of the other player.

(AD) For every set S of infinite sequences of natural numbers, one of the players of the game G_S has a winning strategy (i.e. the game G_S is *determined*).

The trouble with (AD) is that it is not obviously true. A proposed axiom should be immediately acceptable as intuitively evident, and (AD) does not meet this requirement. Its significance lies in its consequences, in relation to (AC).

Theorem 5.27

(i) (AC) implies that (AD) is false.

(ii) (AD) implies that (AC) is false. (This is equivalent to (i).)

(iii) (AD) implies the following restricted version of (CAC): Every non-empty countable collection of non-empty sets (each of which has cardinal number $\leqslant \aleph$) has a choice function.

(iv) (AD) implies that every set of real numbers is Lebesgue measurable.

(v) (AD) implies that every set of real numbers either is countable or has cardinal number $\aleph$, thus deciding the continuum hypothesis. (See Chapter 2.)

(vi) (AD) implies that the set of real numbers cannot be well-ordered.

► It should be emphasised that we make no claims regarding the truth or the acceptability of (AD). Theorem 5.27 merely expresses logical relationships between various principles. Indeed, (AD) is definitely still under suspicion since it is not known whether it is consistent with standard set theory, i.e. whether any contradictions are derivable from it together with the Zermelo–Fraenkel axioms. If it were inconsistent with ZF, of course, it would be valueless as an additional axiom. Theorem 5.27 would remain true, but it would be vacuous.

Let us now conclude the chapter with a list of equivalents and consequences of the axiom of choice, some of which have already been mentioned. These come from various branches of mathematics and some may be familiar in other contexts. We omit proofs. The interested reader is referred, for proofs and other similar results, to the books by Jech and by Rubin & Rubin.

Theorem 5.28
The following principles are all equivalent to (AC).
  (i) Zorn's lemma (Theorem 5.17).
 (ii) Kuratowski's lemma: given any ordered set X, the set of chains in X contains a maximal element (under $\subseteq$). (Exercise 6 on page 183.)
(iii) The well-ordering theorem (Theorem 5.17).
 (iv) For every infinite set A, $A \times A \sim A$ (Theorem (5.15).
  (v) For every pair of sets X and Y, either $X \leqslant Y$ or $Y \leqslant X$ (Theorem 5.13).
 (vi) The maximal ideal theorem for lattices (Theorem 3.24).
(vii) Tychonoff's theorem: the product of a family of compact topological spaces is compact in the product topology.

Theorem 5.29
The following principles are all equivalent to the Boolean prime ideal theorem and, consequently, by Theorem 4.31, are weaker than (AC).
  (i) The Stone representation theorem (Theorem 3.20).
 (ii) The completeness theorems for propositional and predicate logic.
(iii) The compactness theorems for propositional and predicate logic.
 (iv) Tychonoff's theorem, restricted to Hausdorff spaces.

Theorem 5.30
The following principles are all consequences of the axiom of choice.

(i) Every vector space has a basis (Theorem 5.11).
(ii) Every commutative ring contains a maximal ideal.
(iii) The Hahn–Banach theorem: any linear functional on a subspace of a given vector space can be extended to a linear functional on the whole space.
(iv) The Nielsen–Schreier theorem: every subgroup of a free group is free.
(v) For any field F, the algebraic closure of F exists and is unique up to isomorphism.

Exercises

1. Supposing only that for any set X, either $X \leqslant \mathbb{N}$ or $\mathbb{N} \leqslant X$, deduce that every non-empty countable set of non-empty finite sets has a choice function. (It can be shown that (CAC) is not a consequence of this supposition.)
2. Prove that (CAC) implies that every infinite set is Dedekind infinite.
3. Prove that (CAC) implies that the two definitions of limit point in $\mathbb{R}$ (given on page 187) are equivalent.
4. Show that (AC) implies (DC).
5. Prove that (DC) implies (CAC).
6. Very often the following is given as a necessary and sufficient condition for a totally ordered set X to be well-ordered: X contains no infinite descending chain. Prove (using (DC)) that a totally ordered set X is well-ordered if and only if it contains no infinite descending chain.
7. Prove that (AD) implies that every non-empty countable set of non-empty sets of real numbers has a choice function. (Hint: since $\mathbb{R} \sim P(\mathbb{N})$ we may consider subsets of $P(\mathbb{N})$ rather than sets of real numbers.)

Further reading

Barwise [1] This huge volume contains articles on all of the current areas of research in mathematical logic, written for the non-logician. Some of the articles are very technical, but the ones on axiomatic set theory and on the axiom of choice are quite readable.
Enderton [9] See page 162.
Fraenkel, Bar-Hillel & Levy [10] See page 162.
van Heijenoort [14] See page 162.
Jech [16] The definitive work on the axiom of choice, but quite advanced, and mostly requiring knowledge of mathematical logic.
Kuratowski & Mostowski [18] See page 107.
Rubin & Rubin [20] A comprehensive list of equivalents of the axiom of choice.
Sierpinski [22] See page 81.

6

ORDINAL AND CARDINAL NUMBERS

Summary

The first section contains the definition and properties of von Neumann ordinal numbers. It is proved that every well-ordered set is isomorphic to a unique ordinal number. There is a discussion of uncountable ordinals. The second section describes in detail the process of definition of a function or sequence by transfinite induction, through the transfinite recursion theorem. Lastly, cardinal numbers are defined as alephs (initial ordinals), and some properties are derived, including properties of the arithmetic operations on cardinal numbers. There is a brief discussion of the generalised continuum hypothesis and of large cardinals and some of their properties.

Chapters 2 and 3 are prerequisites for this chapter. It is useful, but not essential, also to have read Chapters 4 and 5.

6.1 **Well-ordered sets and ordinal numbers**

We have already mentioned the idea of ordinal number, but it has so far been only an imprecise and informal notion. It has been 'something' which order isomorphic well-ordered sets have in common. We have also seen that ordinal numbers are associated with our so-called generalised counting procedures. In detailed study of the foundations of mathematics, indeed in the demonstrations of some·of the results which we have already mentioned without proof in earlier chapters, ordinal numbers play an essential part, and it is necessary to be more explicit about what they are. Nevertheless, in a sense what they are is less important than the properties that they have. This was the case also with natural numbers, and the same point was made when we gave our formal definition in Chapter 4. Moreover, our definition of ordinal number will be based largely on similar ideas.

Given a well-ordered set, we can conceive of the collection of all well-ordered sets which are order isomorphic to our given set. Under the definition which we shall give, an ordinal number will be a well-ordered set of a particular kind, defined in such a way that each collection such as the above will contain precisely one ordinal number. Thus each well-ordered set will be order isomorphic to a unique ordinal number.

Recall now the relationship between our generalised counting procedures and ordinal numbers. This meant that an ordinal number must have a least element, and each element of an ordinal number must have a successor. The simple trick (invented by von Neumann) in the definition of ordinal numbers is to use the property we discovered about natural numbers (as formally defined in Chapter 4), namely that each natural number *is* the set of all of its predecessors. (See Theorem 4.19(v).)

Definition

An *ordinal number* is a well-ordered set in which each element is equal to the set of all its predecessors. In other words, a set X, well-ordered by R, is an ordinal number if for each $x \in X$,

$$x = \{y \in X : yRx \ \& \ y \neq x\}.$$

Example

In Chapter 4, we defined natural numbers in a formal way and showed that these abstract numbers had the right properties (as described in Chapter 1). Amongst these properties is that the set ω is well-ordered by the relation $\leq$, but further, due to the way in which the abstract numbers are defined, we have, for each $n \in \omega$,

$$n = \{m \in n : m \leq n \ \& \ m \neq n\}.$$

This is the result of Theorem 4.19(v). Thus each natural number (being a subset of ω, and hence well-ordered by $\leq$) satisfies the definition of ordinal number. In detail, let $n \in \omega$ and let $k \in n$. Then $k = \{m \in k : m \leq k \ \& \ m \neq k\} = \{m \in n : m \leq k \ \& \ m \neq k\}$, as required.

A similar argument shows that ω itself, ordered by magnitude, satisfies the definition of ordinal number.

▶ The rest of this section is devoted to showing that the definition above fulfils the required purpose, namely, that every well-ordered set is order isomorphic to a unique ordinal number, and to deriving some properties of ordinal numbers in order to build up an intuitive picture of what they are and how they behave. These two objectives will be pursued together, since each will assist the other.

First let us establish some notation and terminology.

Definitions

(i) Let X be a set, well-ordered by the relation R. The *initial segment* of X determined by an element a of X is the set $\{y \in X : yRa \ \& \ y \neq a\}$. This set is denoted by X_a. It is well-ordered by the restriction of R.

(ii) We shall use the words *isomorphism* and *isomorphic* to mean order isomorphism and order isomorphic. This will not cause confusion. Also, if sets X and Y are well-ordered by the relations R and S respectively, we write $(X, R) \simeq (Y, S)$, or, if the orderings are well understood, just $X \simeq Y$. (Note: In some textbooks the word *similar* is used to mean order isomorphic, when applied to well-ordered sets. We do not follow this practice, since 'similar' is over-used elsewhere in mathematics in any case.)

▶ Next we derive some results about well-ordered sets and isomorphisms.

Remarks 6.1

Let X be a set, well-ordered by the relation R.

(a) Let $a \in X$ and let $b \in X_a$. Then $(X_a)_b = X_b$. This may make sense if we consider the set X represented by a line (see Fig. 6.1), where xRy if and only if x is to the left of (or equal to) y. X_a is the set of points to the left of a. $(X_a)_b$ is the set of points of X_a to the left of b, which is clearly just X_b.

Fig. 6.1

(b) Let $a \in X$, $b \in X$. Then aRb if and only if $X_a \subseteq X_b$. Again, reference to a diagram makes this clear.

(c) Let $a \in X$. Then a is the least element in $X \backslash X_a$. Note, in particular, that $a \notin X_a$.

Theorem 6.2

Let X be a set, well-ordered by the relation R, and let Y be a subset of X such that for each $b \in Y$, $X_b \subseteq Y$. Then either $Y = X$ or Y is an initial segment of X.

Proof

Suppose that $Y \neq X$. Let u be the least element in $X \backslash Y$. We show that $Y = X_u$. First, let $b \in Y$ and $b \notin X_u$. Then $u \in X_b$, so by our assumption about Y, we have $u \in Y$, which is a contradiction. Thus $Y \subseteq X_u$. Conversely, $X_u \subseteq Y$ since u is the least element of X not in Y.

Theorem 6.3

Let X and Y be sets, well-ordered by the relations R and S respectively.

(i) If $f : X \to X$ is an order-preserving injection then for every $a \in X$, we have $aRf(a)$.

(ii) If $f : X \to X$ is an isomorphism, then f is the identity function.

(iii) If $f : X \to Y$ and $g : X \to Y$ are isomorphisms then $f = g$.

(iv) X cannot be isomorphic to an initial segment of itself.

Proof (see Exercise 19 on page 96)

(i) Suppose that $a \in X$ and not $aRf(a)$. Then $a \neq f(a)$ and $f(a)Ra$ (since R is a total order). Since f is an injection and order-preserving, we must have $f(a) \neq f(f(a))$ and $f(f(a))Rf(a)$. Similarly, $f^2(a) \neq f^3(a)$ and $f^3(a)Rf^2(a)$. We thus generate the set $\{f^n(a) : n \in \mathbb{N}\}$ which is a non-empty subset of X with no least element. This cannot exist, since R is a well-ordering. The result follows.

(ii) Let $f : X \to X$ be an isomorphism. Then f^{-1} is also an isomorphism. By (i) above, then, for each $a \in X$ we have $aRf(a)$ and $f(a)Rf^{-1}(f(a))$, i.e. $f(a)Ra$. Consequently $a = f(a)$, and f must be the identity function.

(iii) Let $f : X \to Y$ and $g : X \to Y$ be isomorphisms. Then $g^{-1}f : X \to X$ is an isomorphism, so by (ii), $g^{-1}f$ is the identity function on X, and hence $g = f$.

(iv) Suppose that $b \in X$ and $f : X \to X_b$ is an isomorphism. Then f is an order-preserving injection from X into X. By (i), for each $a \in X$ we have $aRf(a)$. In particular, $bRf(b)$. But $f(b)$ must lie in X_b and $X_b = \{y \in X : yRb \ \& \ y \neq b\}$. Hence, $f(b)Rb$ and $f(b) \neq b$. This contradicts $bRf(b)$, and the required result follows.

Theorem 6.4

Given any two well-ordered sets, either they are isomorphic or one is isomorphic to an initial segment of the other.

Proof

Let X and Y be sets, well-ordered by the relations R and S respectively, and suppose that X and Y are not isomorphic. If either X or Y is empty then the result is trivial, so suppose that both sets are non-empty. Then X and Y have least elements, say x_0 and y_0. We shall consider initial segments of X and of Y. Trivially, X_{x_0} is empty and Y_{y_0} is empty, so x_0 and y_0 determine isomorphic initial segments. Now let $a \in X$ and suppose that X_a is isomorphic to an initial segment Y_b of Y. By Theorem 6.3(iv) such b (if it exists) is unique. So we can define a subset A of X and a function $\phi : A \to Y$ thus:

$$A = \{a \in X : \text{there exists } b \in Y \text{ with } X_a \simeq Y_b\},$$

$$\phi(a) = b \text{ if } a \in A \text{ and } X_a \simeq Y_b.$$

We shall show that ϕ is one-one and order-preserving and that either $A = X$ and $\phi(A)$ is an initial segment of Y or A is an initial segment of X and $\phi(A) = Y$.

Let $a \in A$ and let $x \in X$ with xRa (and $x \neq a$). We have an isomorphism $f : X_a \simeq Y_b$ for some $b \in Y$. Now $x \in X_a$, so $f(x) \in Y_b$ and certainly the restriction of f is an isomorphism from X_x to $Y_{f(x)}$. Hence, $x \in A$. By Theorem 6.2, then, either $A = X$ or A is an initial segment of X. A similar argument shows that either $\phi(A) = Y$ or $\phi(A)$ is an initial segment of Y.

Now let $a, c \in A$ with aRc. We show that $\phi(a)S\phi(c)$. If $a = c$ this is trivial, so suppose that $a \neq c$, so that $a \in X_c$. Since $c \in A$, there is an isomorphism $g : X_c \to Y_{\phi(c)}$. But $X_a \subseteq X_c$, so the restriction of g to X_a is an isomorphism between X_a and $Y_{g(a)}$ with $g(a) \in Y_{\phi(c)}$, i.e. $g(a)S\phi(c)$. However, $g(a) = \phi(a)$ since $X_a \simeq Y_{\phi(a)}$ also. Hence $\phi(a)S\phi(c)$, as required, and ϕ is order-preserving.

To complete the proof, let us consider first the case when $A \neq X$ and $\phi(A) \neq Y$. Let c be the least element of $X \backslash A$ and let d be the least element of $Y \backslash \phi(A)$. Then $A = X_c$ and $\phi(A) = Y_d$. Also, ϕ is an isomorphism between X_c and Y_d, and so $c \in A$, which is a contradiction. This situation therefore cannot arise, so we must have either $A = X$ or $\phi(A) = Y$. It follows that either ϕ is an isomorphism from X to an initial segment of Y or ϕ^{-1} is an isomorphism from Y to an initial segment of X.

▶ It is of interest to compare Theorem 6.4 with Theorem 5.13 (given any two sets, either they are equinumerous or one strictly dominates the other) and to notice that the axiom of choice was required there but is not required here. Observe that Theorem 6.4 implies a special case of Theorem 5.13, namely: given any two *well-ordered* sets, either they are equinumerous or one dominates the other, and that this does not require (AC) for its proof.

Ordinal numbers are well-ordered sets of a special kind. The general results above, when applied to ordinal numbers, yield some interesting and important properties. For convenience we abbreviate 'ordinal number' by 'ordinal'. We adopt the standard practice and denote ordinals by lower case Greek letters.

Theorem 6.5
(i) An initial segment of an ordinal is an ordinal. Equivalently, every element of an ordinal is an ordinal.
(ii) The order relation on an ordinal is always $\subseteq$.
(iii) If two ordinals are isomorphic then they are equal.
(iv) If α and β are ordinals then one of the following holds:

$$\alpha = \beta, \; \alpha \in \beta, \; \beta \in \alpha.$$

Proof
(i) Let α be an ordinal and let $x \in \alpha$. Then $x = \alpha_x$, the initial segment of α determined by x. Now if $y \in \alpha_x$ then $y \in \alpha$, so $y = \alpha_y$ also. But $\alpha_y = (\alpha_x)_y$, by Remark 6.1(a), and $(\alpha_x)_y = x_y$. Hence,

$y = x_y$, for every $y \in x$.

Therefore x satisfies the definition of ordinal number.

We have shown simultaneously that every element x of the ordinal α is an ordinal and every initial segment α_x of the ordinal α is an ordinal.

(ii) Let α be an ordinal, and suppose that the order relation on α is denoted by R. We show that for $x, y \in \alpha$, xRy if and only if $x \subseteq y$. By Remark 6.1(b), for $x, y \in \alpha$ we have xRy if and only if $\alpha_x \subseteq \alpha_y$. But $x = \alpha_x$ and $y = \alpha_y$ since α is an ordinal. Consequently, xRy if and only if $x \subseteq y$.

(iii) Let α and β be ordinals, and let $f : \alpha \to \beta$ be an isomorphism. Suppose that f is not the identity function on α. Then the set

$\{x \in \alpha : f(x) \neq x\}$ is not empty, and so contains a least element (under the ordering in α), say x_0. For $x \subseteq x_0$ in α, then, with $x \neq x_0$, we have $f(x) = x$. That is, f is the identity function between α_{x_0} and $\beta_{f(x_0)}$. It follows that $\alpha_{x_0} = \beta_{f(x_0)}$. But α and β are ordinals, so $\alpha_{x_0} = x_0$ and $\beta_{f(x_0)} = f(x_0)$. We therefore have $x_0 = f(x_0)$, which contradicts our definition of x_0, and so f must be the identity function on α. Hence $\alpha = \beta$.

(iv) Let α and β be ordinals. Then by Theorem 6.4 one of the following holds: α is isomorphic to β, α is isomorphic to an initial segment of β, or β is isomorphic to an initial segment of α. By parts (i) and (iii) then, one of the following holds: $\alpha = \beta$, α is equal to an element of β, or β is equal to an element of α. This is the required result.

Corollary 6.6

If α and β are distinct ordinals with $\alpha \subseteq \beta$, then $\alpha \in \beta$.

Proof

By (iv) of the theorem, we have either $\alpha \in \beta$ or $\beta \in \alpha$. If $\beta \in \alpha$ then since α is an ordinal, $\beta = \alpha_\beta \subseteq \alpha$. Together with $\alpha \subseteq \beta$, this yields $\alpha = \beta$, which contradicts the hypothesis. Hence we have $\alpha \in \beta$.

▶ Recall that the natural numbers as defined in Chapter 4 are ordinals. Indeed, any finite set which is an ordinal number is a natural number. This is because, as observed on page 174, a finite set can be well-ordered in essentially only one way, and consequently if α is an ordinal with n elements then α is isomorphic to the ordinal number n, and hence equal to it.

Let us consider the results of Theorem 6.5 in relation to the natural numbers. For each $n \in \omega$, we know that $n^+ = \{0, 1, \ldots, n\}$. It is an easy consequence of this that $m \leq n$ if and only if $m \subseteq n$. Also that $m < n$ if and only if $m \in n$, and $m \in n$ if and only if m is an initial segment of n. As for (iv), we have, for natural numbers m and n, $m = n$ or $m \in n$ or $n \in m$. This could have been proved directly in Chapter 4, but it now follows from the more general result.

In our previous discussion we described ways of generating larger ordinal numbers from given ones. The first was through successors and the second was through unions. Now we shall consider these in general.

Theorem 6.7

If α is an ordinal, then $\alpha \cup \{\alpha\}$ is an ordinal (the *successor* of α) denoted by α^+, as before.

Proof

Let α be an ordinal. Then α is well-ordered by $\subseteq$. We show first that $\alpha \cup \{\alpha\}$ is totally ordered by $\subseteq$. Let $x, y \in \alpha \cup \{\alpha\}$. If both x and y belong to α, then either $x \subseteq y$ or $y \subseteq x$ since α is totally ordered by $\subseteq$. If both $x = \alpha$ and $y = \alpha$ then $x \subseteq y$, trivially. Let $x \in \alpha$ and $y = \alpha$, then α is an ordinal, so $x = \alpha_x \subseteq \alpha = y$, so $x \subseteq y$. Hence, $\alpha \cup \{\alpha\}$ is totally ordered by $\subseteq$. Now let X be a non-empty subset of $\alpha \cup \{\alpha\}$. If $X = \{\alpha\}$ then X trivially contains a least element. If $X \cap \alpha \neq \emptyset$ then $X \cap \alpha$ contains a least element, since α is well-ordered, and this element is least in X, since we have $x \subseteq \alpha$ for each $x \in \alpha$ (by the above argument). Thus $\alpha \cup \{\alpha\}$ is well-ordered by $\subseteq$. Last, to show that $\alpha \cup \{\alpha\}$ is an ordinal, let $x \in \alpha \cup \{\alpha\}$. If $x = \alpha$ then $(\alpha \cup \{\alpha\})_x = \alpha = x$. If $x \neq \alpha$, so that $x \in \alpha$, then $(\alpha \cup \{\alpha\})_x = \alpha_x = x$. This completes the proof.

Theorem 6.8

(i) Any set of ordinal numbers is well-ordered by $\subseteq$.

(ii) The union of any set of ordinal numbers is an ordinal number.

Proof

(i) Let X be a set whose elements are ordinals. X is ordered by $\subseteq$. Let $\alpha, \beta \in X$, with $\alpha \neq \beta$. Then either $\alpha \in \beta$ or $\beta \in \alpha$, by Theorem 6.5(iv). If $\alpha \in \beta$ then by the definition of ordinal number, $\alpha = \beta_\alpha \subseteq \beta$. Similarly, if $\beta \in \alpha$ then $\beta \subseteq \alpha$. Thus $\subseteq$ is a total order on X. Finally, let Y be a non-empty subset of X and let γ be some chosen element of Y. If γ is least in Y then there is nothing further to prove, so suppose that γ is not least in Y. If $\delta \in Y$ and $\delta \subseteq \gamma (\delta \neq \gamma)$ then δ is an ordinal and $\delta \in \gamma$, by Corollary 6.6. Now $Y \cap \gamma$ is a non-empty subset of γ, and γ is well-ordered by $\subseteq$, so $Y \cap \gamma$ contains a least element, which is necessarily least in Y also. Thus X is well-ordered by $\subseteq$.

(ii) Let X be a set whose elements are ordinals. First we show that the elements of $\bigcup X$ are ordinals also. Let $x \in \bigcup X$. Then $x \in \alpha$ for some ordinal $\alpha \in X$. But Theorem 6.5(i) then implies that x is an ordinal. Next, by part (i) above, $\bigcup X$ is well-ordered by $\subseteq$. It remains to show that for each $x \in \bigcup X$, $(\bigcup X)_x = x$, in order to verify that $\bigcup X$ is an ordinal. Let $x \in \bigcup X$, so that $x \in \alpha$, for some $\alpha \in X$. Now let $y \in x$. Since x is an ordinal, y is an initial segment of x. Also, x is an initial segment of α, so it follows that y is an initial segment of α, and consequently $y \in \alpha$.

Hence, $y \in \bigcup X$, and we have shown that $x \subseteq \bigcup X$. Now

$$(\bigcup X)_x = \{z \in \bigcup X : z \subseteq x \text{ and } z \neq x\}$$
$$= \{z \in \bigcup X : z \in x\},$$

since $\bigcup X$ is a set of ordinals, using Corollary 6.6. Thus $(\bigcup X)_x = (\bigcup X) \cap x = x$, since $x \subseteq \bigcup X$. The proof is now complete.

▶ The union of a set of ordinals, besides being an ordinal, has an important further property, which will help in our intuitive picture of of what the system of ordinals is like.

Corollary 6.9

Let X be a set of ordinals. Then $\bigcup X$ is the smallest ordinal which is larger than or equal to every ordinal in X. In other words, $\bigcup X$ is the least upper bound for X in the system of ordinals.

Proof

We know that $\bigcup X$ is an ordinal. Let $\alpha \in X$. Then $\alpha \subseteq \bigcup X$, by the definition of the union. Hence $\bigcup X$ is an upper bound for X. Now let β be an upper bound for X, i.e. suppose that $\alpha \subseteq \beta$ for every $\alpha \in X$. We must show that $\bigcup X \subseteq \beta$, so let $\gamma \in \bigcup X$. Then $\gamma \in \alpha$ for some $\alpha \in X$. But then $\gamma \in \beta$, since $\alpha \subseteq \beta$. Hence, $\bigcup X \subseteq \beta$, as required.

Example 6.10

We have seen that each natural number n is an ordinal. Also ω is an ordinal. Observe that $\bigcup \omega = \omega$, so taking unions gives nothing new. Also, if $\{n_1, \ldots, n_k\}$ is a set of natural numbers in increasing order, $\bigcup \{n_1, \ldots, n_k\} = n_k$. The obvious way to generate new ordinals is through successors, starting with ω; ω^+, $(\omega^+)^+$, $((\omega^+)^+)^+$, ... are all ordinals. It is convenient to denote these by $\omega + 1$, $\omega + 2$, $\omega + 3$, ... For the moment, this notation has no connotations with respect to an addition operation. That will come later. In fact $\{\omega + n : n \in \omega\}$ is a *set* of ordinals, as can be demonstrated in ZF using the generalised recursion theorem (4.20). (See Exercise 18 on page 145.) $\bigcup \{\omega + n : n \in \omega\}$ is therefore an ordinal, by Theorem 6.8(ii). It is denoted by $\omega 2$. Another sequence: $\omega 2$, $(\omega 2)^+$, $((\omega 2)^+)^+$, ... may be generated, and its union taken, to obtain the ordinal $\omega 3$. Continuing thus, we obtain a set of ordinals $\{\omega n : n \in \omega\}$. Its union is an ordinal, denoted by ω^2. This process is clearly endless; moreover, it is endlessly endless. For, given any infinite increasing sequence of ordinals α_1, α_2, α_3, . . . , say, we can construct a larger ordinal, namely, the union $\bigcup\{\alpha_i : i \in \mathbb{Z}^+\}$, and from that we can start another endless sequence.

Definition
Non-zero ordinals may be of either of two kinds. An ordinal α is a *successor ordinal* if $\alpha = \beta^+$, for some ordinal β. Otherwise α is a *limit ordinal*. A limit ordinal does not have an immediate predecessor.

Example 6.11
From the previous example: ω, $\omega 2$, $\omega 3$, ω^2 are all limit ordinals. These are the ones which are generated as unions of endless sequences.

▶ We should note here that the processes of taking successors and unions, starting with countable ordinals, will yield only countable ordinals, no matter how far they are taken (since a countable union of countable sets is countable – Theorem 5.21). At this stage we have no hard information about whether such things as uncountable ordinals exist, but observe that such an object would be an uncountable well-ordered set. As we observed before, well-orderings of uncountable sets are at best difficult to describe. Note, however, that given an uncountable ordinal α, we can construct from it a hierarchy of larger ordinals by means of successors and unions, and all members of this hierarchy will be equinumerous with α.

A note of caution before we proceed, however.

Theorem 6.12
There is no set of all ordinals.

Proof
Suppose that X is the set of all ordinals. Then $\bigcup X$ is an ordinal, by Theorem 6.8(ii). We derive a contradiction by finding an ordinal which cannot belong to X, so note first that it is possible that $\bigcup X \in X$ (as an exercise, the reader may ponder the circumstances which would cause this). Let $\alpha = (\bigcup X)^+$ $(=\bigcup X \cup \{\bigcup X\})$. Now let $\gamma \in X$. Then $\gamma \subseteq \bigcup X$, so either $\gamma \in \bigcup X$ or $\gamma = \bigcup X$, by Corollary 6.6. If $\gamma \in \bigcup X$ then γ is an initial segment of $\bigcup X$ and so of α. If $\gamma = \bigcup X$ then γ is an initial segment of $(\bigcup X)^+$ i.e. of α. In either case $\gamma \neq \alpha$. Hence, α is an ordinal number and $\alpha \notin X$. This contradicts our original assumption, so such a set X cannot exist.

▶ This theorem, which we are now able to take in our stride, represents the resolution of what was once regarded as a substantial logical paradox, the Burali–Forti paradox. This was first formulated in 1897, and briefly it is as follows. Given any ordinal number, there is a larger ordinal number. But there cannot be an ordinal number which is larger than the ordinal number determined by the set of all ordinal numbers.

In ZF set theory, the collection of all ordinals is not an object which can be mentioned. It is analogous in this sense to the collection of all sets. However, in VNB set/class theory, each ordinal number is a set and the collection of all ordinals, again like the collection of all sets, is a proper class, and as such can be discussed within the system. The interested reader may care to ponder how 'x is an ordinal' may be expressed as a well-formed formula, so that the comprehension axiom (VNB9) may be applied.

Theorem 6.12 provides the basis for the main result about ordinals, mentioned earlier, which is the following.

Theorem 6.13

Given any set X, well-ordered by the relation R, there is a unique ordinal number which is isomorphic to it. It is called the ordinal number of (X, R) and is denoted by ord (X, R).

Proof

If such an ordinal exists, then it is unique by Theorem 6.5(iii). We have only to demonstrate existence, therefore. Given a set X, for any ordinal α, by Theorem 6.4 one of the following holds: X is isomorphic to α, X is isomorphic to an initial segment of α, or α is isomorphic to an initial segment of X. If X is isomorphic to an initial segment of α, then X is isomorphic to an ordinal number, by Theorem 6.5(i). Suppose that (X, R) is not isomorphic to an ordinal number. From the above discussion, then, it follows that for every ordinal number α, α is isomorphic to an initial segment of X. Now we deduce that there is a set of all ordinals, using the replacement axiom (ZF7). Let

$$D = \{Y \in P(X): Y \text{ is an initial segment of } X \text{ and } Y$$
$$\text{is isomorphic to some ordinal number}\},$$

and let $\mathscr{A}(x, y)$ be the statement: $x \in D$ and y is the ordinal number isomorphic to x. By the replacement axiom,

$$\{\alpha : (\exists x)\mathscr{A}(x, \alpha)\}$$

is a set. But, as we have seen, this set contains all the ordinals, and this is impossible. This completes the proof.

Notation

In view of our results about the elements of an ordinal and the order relation on an ordinal, we know that every ordinal is the set of all smaller ordinals, so that, for ordinals α and β,

$$\alpha \in \beta \text{ if and only if } \alpha \subseteq \beta \text{ and } \alpha \neq \beta.$$

We may therefore use $\subseteq$ or $\in$ to describe the order relation on an ordinal. It will be convenient to return to use of the standard inequality sign $\leq$, extending its use from the set ω as defined in Chapter 4 to ordinals in general. We therefore write $\alpha \leq \beta$, for ordinals α and β, if $\alpha \subseteq \beta$, i.e. if $\alpha \in \beta$ or $\alpha = \beta$, and we write $\alpha < \beta$ if $\alpha \subseteq \beta$ and $\alpha \neq \beta$, i.e. if $\alpha \in \beta$.

▶ Let us close the section with some further comments on uncountable ordinals. We know that there are uncountable sets; $\mathbb{R}$ and $P(\mathbb{N})$ are examples. We do not know of any well-ordering of an uncountable set. However, under the assumption of the axiom of choice we know that given any set X, there is a relation on X which is a well-ordering of X (Theorem 5.17), and in particular, $\mathbb{R}$ can be well-ordered. By Theorem 6.13, then, there is an ordinal number which is equinumerous with $\mathbb{R}$. This ordinal is certainly uncountable. There will be other ordinals which are equinumerous with this one, as noted earlier, but we can see as follows that there must be a least uncountable ordinal. Let α be some uncountable ordinal, and suppose that α is not the least such. All smaller ordinals are elements of α, so the least uncountable ordinal (if it exists) will belong to α. Consider the set $\{\beta \in \alpha : \beta \text{ is an uncountable ordinal}\}$. This set is not empty, and it is a subset of the well-ordered set α, so it has a least element.

Notation ·

The least uncountable ordinal is usually denoted by ω_1.

▶ Notice that since every ordinal is the set of all smaller ordinals, ω_1 is the set of all countable ordinals. But note also that we did not just define ω_1 as the set of all countable ordinals. We require to be sure that all countable ordinals actually constitute a *set*. One way of doing this is to show (as we did above) that there is an ordinal larger than ω_1, and in the above argument we have to appeal to the axiom of choice to do so. There is another approach, however, by means of which we can dispense with the axiom of choice in the demonstration of the existence of ω_1. We show, using the axioms of ZF only, that the collection of all countable ordinals is a set.

The replacement axiom (ZF7) guarantees the existence of the set of all countable ordinals as follows. We take the formula $\mathscr{F}(R, \alpha)$ to be:

'R is a relation which well-orders a subset A of ω,
α is an ordinal and $\alpha = \text{ord}\,(A, R)$.'

This is clearly not a well-formed formula of ZF, but by methods we have already seen, it can be translated into a formula of the required kind. It is clear that $\mathscr{F}$ determines a function. By axiom (ZF6), there is a set

$$W = \{R \in P(\omega \times \omega) : R \text{ is a well-ordering of a subset of } \omega\}.$$

By (ZF7) then, the collection of all ordinal numbers α such that $\mathscr{F}(R, \alpha)$ holds for some well-ordering R of a subset of ω, is a set. This is the set of all countable ordinals. For if α is as described above then $\alpha = \text{ord}\,(A, R)$ for some $A \subseteq \omega$, so $\alpha \sim A$ and α is countable. Conversely, if α is a countable ordinal then either α is finite or $\alpha \sim \omega$. If α is finite then α is itself a subset of ω, and $\alpha = \text{ord}\,(\alpha, \subseteq)$. If $\alpha \sim \omega$ then there is a bijection $f : \omega \to \alpha$, and we can define a well-ordering R on ω by $(m, n) \in R$ if and only if $f(m) \subseteq f(n)$ in α. Then $\alpha = \text{ord}\,(\omega, R)$ as required.

Finally, let us note that the existence of ω_1, shown without recourse to the axiom of choice, does not bring us any nearer to a description or construction of a well-ordering of $\mathbb{R}$, since without (AC) we know of no relationship between ω_1 and $\mathbb{R}$ apart from the fact that both are uncountable sets.

Exercises

1. Prove that ω is an ordinal number.
2. Let α and β be distinct ordinal numbers. Is $\alpha \backslash \beta$ an ordinal number?
3. (i) Is $(P(\omega), \subseteq)$ an ordinal number?
   (ii) Is $(\omega \times \omega, \subseteq)$ an ordinal number?
4. Let α and β be distinct ordinal numbers, and let $f : \alpha \to \beta$ be an order-preserving injection. Prove that $\alpha \in \beta$. Is the image of f necessarily an initial segment of β? Is it necessarily an ordinal number?
5. Let X be a non-empty set of ordinals. Prove that $\bigcap X$ is an ordinal number, and $\bigcap X$ is the least element of X.

6. Prove that for any ordinal α, $\bigcup \alpha^+ = \alpha$. Prove that if α is a limit ordinal then $\bigcup \alpha = \alpha$.

7. Let X be a set of ordinals. Show that $X \subseteq \bigcup X$.

8. Let X be a set of ordinals and let Y be a subset of X which is cofinal in X (i.e. given any $\alpha \in X$ there is $\beta \in Y$ with $\alpha \leq \beta$). Prove that $\bigcup X = \bigcup Y$.

9. Let X be an infinite set of ordinals. In what circumstances is $\bigcup X$ a limit ordinal, and in what circumstances is $\bigcup X$ a successor ordinal?

10. Show that, given any set X of ordinals, there is an ordinal which is greater than every element of X.

11. (i) Show that, if α is an ordinal number and $X \subseteq \alpha$, then ord $(X, \subseteq) \leq \alpha$.
    (ii) Let (A, R) be a well-ordered set, and let $X \subseteq A$. Prove that ord $(X, R) \leq$ ord (A, R).

12. Show, by finding an appropriate formula of VNB and applying axiom (VNB9) to it, that in VNB there exists a class of all ordinals.

6.2 Transfinite recursion and ordinal arithmetic

Given a set X, it may happen that X is equinumerous with some ordinal number α. A bijection $f: \alpha \to X$ imposes a well-ordering R on X by the rule: for $a, b \in X$, $(a, b) \in R$ if and only if $f^{-1}(a) \leq f^{-1}(b)$. Note that $f^{-1}(a)$ and $f^{-1}(b)$ are elements of the ordinal number α, and that $f^{-1}(a)$ and $f^{-1}(b)$ are consequently themselves ordinals. Such a bijection is in effect what we have referred to earlier as a generalised counting procedure.

Examples 6.14

The set ω may be enumerated by various generalised counting procedures, as we have seen.

(a) $0, 1, 2, 3, \ldots$ corresponds to the bijection $f: \omega \to \omega$ such that $f(n) = n$.

(b) $1, 2, 3, \ldots, 0$ corresponds to the bijection $g: \omega^+ \to \omega$ given by $g(n) = n + 1$, for $n \in \omega$, and $g(\omega) = 0$.

(c) $0, 2, 4, 6, \ldots, 1, 3, 5, \ldots$ corresponds to the bijection $h: \omega 2 \to \omega$ given by $h(n) = 2n$, for $n \in \omega$, and $h(\omega + n) = 2n + 1$, for $n \in \omega$.

▶ Another way of describing this sort of object is as a transfinite sequence.

Definition

A *transfinite sequence* is a function (often, but not necessarily, an injection) whose domain is an ordinal number.

In this situation, the ordinal number acts as a kind of index set. One of the areas of application of ordinals in mathematics is through the representation of sets as transfinite sequences indexed by ordinal numbers. Observe that the axiom of choice is required for justification that *every* set can be so represented. In this area there are two important tools: the principle of transfinite induction and the transfinite recursion theorem. The former we have already come across (in Chapter 5). The latter is the basis for construction of transfinite sequences.

Theorem 6.15 (principle of transfinite induction)
Let X be a non-empty set, well-ordered by the relation R, and let A be a non-empty subset of X satisfying:

if $X_a \subseteq A$, then $a \in A$,

for each $a \in X$. Then $A = X$.

Proof
Left as an exercise in Chapter 5 (see Exercise 3 on page 183).

Corollary 6.16 (transfinite induction on ordinals)
Let α be a non-zero ordinal and let A be a non-empty subset of α satisfying:

if $\gamma \subseteq A$, then $\gamma \in A$,

for each $\gamma < \alpha$. Then $A = \alpha$.

Proof
This is not difficult if we recall that for ordinals α and γ, if $\gamma < \alpha$ then $\gamma \in \alpha$ and $\alpha_\gamma = \gamma$.

► Ordinal numbers may be successor ordinals or limit ordinals. Thus in order to give a 'definition by induction' of a function f whose domain is to be an ordinal number α we must specify the base of the induction, i.e. give the value of $f(0)$, and we must specify, for each $\beta \in \alpha$, the value of $f(\beta)$ in terms of previous values, i.e. in terms of the values $f(\gamma)$ for $\gamma < \beta$. Thus to specify only each $f(\gamma^+)$ in terms of $f(\gamma)$ will not be sufficient, although it was in our earlier recursion theorem (Theorem 4.15). Values at limit ordinals must also be specified. We illustrate the idea of iterating a construction into the transfinite, and so obtaining a transfinite sequence, by means of an example.

Example 6.17

Let A be a subset of the real line. The derived set A' of A is the set of all limit points of A. Let us define by (so far intuitive) transfinite induction, a transfinite sequence $A_0, A_1, \ldots, A_\omega, \ldots$ by:

$$A_0 = \bar{A} \quad \text{(the topological closure of } A\text{)},$$

$$A_{\gamma^+} = (A_\gamma)', \quad \text{for each ordinal } \gamma,$$

and

$$A_\lambda = \bigcap\{A_\gamma : \gamma < \lambda\}, \quad \text{for each limit ordinal } \lambda.$$

Notice that there are three parts to this definition by induction: the base step and two inductive steps which specify the values at successor ordinals and limit ordinals separately. There is no difficulty about generating the sequence $A_0, A_1, A_2, \ldots$ for finite suffixes. At ω however, we apply the third part to obtain

$$A_\omega = \bigcap\{A_\gamma : \gamma < \omega\} = \bigcap\{A_n : n \in \omega\}.$$

The second part then gives,

$$A_{\omega^+} = (A_\omega)',$$

and so on until we require the third part at $\omega 2$:

$$A_{\omega 2} = \bigcap\{A_\gamma : \gamma < \omega 2\}.$$

This process continues through the sequence of ordinals indefinitely.

► Consideration of examples such as this, and comparison with definitions by induction on ω suggest that this transfinite induction procedure will yield a function defined on all ordinals, that is to say a sequence indexed by the collection of all ordinals. Now there is a formal difficulty about this. We have shown that there is no set of all ordinals, so in ZF set theory there can be no function whose domain is the collection of all ordinals. Thus in ZF set theory there can be no transfinite sequences indexed by the collection of *all* ordinals. The best we can expect is for such sequences to extend up to a particular bounding ordinal.

In VNB set/class theory, the collection of all ordinals is a proper class, and so does actually exist as an object of discussion. In consequence, it is possible to discuss in VNB functions whose domain is this proper class. However, such functions are themselves necessarily proper classes, and proper classes are difficult things to handle. So although this approach has a formal appeal, we shall not follow it, in the hope that we may find a better understanding, independent of these formal constraints.

We therefore shall express the result in two different ways, firstly as a formally expressed theorem, and secondly as a less formal description of the way it works in practice. Understanding of the meaning of the statement of the theorem (and corollary) is more important than following the details of the proofs, so these proofs will be postponed until after some explanatory remarks.

Theorem 6.18 (transfinite recursion theorem)

Let α be an ordinal, let X be a set, and let $\mathscr{S}^\alpha$ be the set of all transfinite sequences of length less than α of elements of X (formally $\mathscr{S}^\alpha = \{f : f$ is a function from β to X, for some $\beta < \alpha\}$). Given any function $h : \mathscr{S}^\alpha \to X$ there exists a unique function (transfinite sequence of length α) $g : \alpha \to X$ such that for all ordinals $\gamma < \alpha$, we have

$$g(\gamma) = h(g|\gamma),$$

where $g|\gamma$, the restriction of g to γ, is the transfinite sequence of all previous values of g (and as such is a member of $\mathscr{S}^\alpha$).

Corollary 6.19

Let $\mathscr{R}$ be a rule which associates a uniquely defined set with any given transfinite sequence (we can think of $\mathscr{R}$ as a formula $\mathscr{R}(f, x)$ involving two variables, which for each transfinite sequence f, holds for precisely one set x, namely, the set to be associated by the rule with the sequence f). Then there is a (unique) similar rule which associates with each ordinal number α a unique set X_α in such a way that for each ordinal α, the set X_α is the result of applying the rule $\mathscr{R}$ to the transfinite sequence $\{X_\beta : \beta < \alpha\}$.

Remarks 6.20

(a) Theorem 6.18 provides for the definition by transfinite induction of a function whose domain is a given ordinal number α, given a function h. This function h represents a construction process by which, given a transfinite sequence of length γ, another element of the set X may be found which is to be the next member of the sequence under construction. Likewise, in Corollary 6.19, the rule $\mathscr{R}$ corresponds to the construction procedure for finding the 'next' member of the sequence at each stage.

(b) Both the theorem and the corollary combine the three separate stages of the induction process (as exemplified in Example 6.17) into one single condition: $g(\gamma) = h(g|\gamma)$ for $\gamma < \alpha$, in the theorem. Let us see how to recover the three from the one.

First, putting $\gamma = 0$ gives

$$g(0) = h(g|0).$$

Now $g|0$ is the empty function, since 0 has no predecessors, and by convention we regard this as an element of $\mathscr{S}^\alpha$. Since h is a given function with domain $\mathscr{S}^\alpha$, then, $g(0)$ has a specified value.

Second, putting $\gamma = \delta^+$ yields

$$g(\delta^+) = h(g|\delta^+).$$

Now $g|\delta^+$ yields the transfinite sequence of values $g(0)$, $g(1)$, ..., $g(\delta)$, so in this case our successor step is more general than that in Example 6.17, where A_{γ^+} depended only on A_γ, the immediately preceding value.

Third, putting $\gamma = \lambda$, a limit ordinal, gives

$$g(\lambda) = h(g|\lambda),$$

which corresponds exactly with the situation in Example 6.17. Here $g(\lambda)$ is specified in terms of the previous values.

Proof

We show first that Corollary 6.19 is a consequence of Theorem 6.18. Let $\mathscr{R}$ be a rule as specified. The rule whose existence is asserted may be described as follows. Given an ordinal α, the restriction of $\mathscr{R}$ to sequences of length less than α^+ yields a function with domain $\mathscr{S}^{\alpha^+}$, which we may denote by h. Applying Theorem 6.18 to the ordinal α^+ and the function h, we obtain a unique function g_α with domain α^+ such that, for every ordinal $\gamma < \alpha^+$,

$$g_\alpha(\gamma) = h(g_\alpha|\gamma).$$

In particular,

$$g_\alpha(\alpha) = h(g_\alpha|\alpha).$$

Now let X_α be the set $g_\alpha(\alpha)$. This can be done for each ordinal α, so to each ordinal α we have associated a set X_α, as required. It remains to show that X_α is the result of applying rule $\mathscr{R}$ to the sequence $\{X_\beta : \beta < \alpha\}$, for each ordinal α. Now $h(g_\alpha|\alpha)$ is the result of applying $\mathscr{R}$ to the sequence $g_\alpha|\alpha$, i.e. $\{g_\alpha(\beta) : \beta < \alpha\}$. We must show, therefore, that for all $\beta < \alpha$, $g_\alpha(\beta) = X_\beta = g_\beta(\beta)$. By Theorem 6.18, g_β is the unique function with the property that $g_\beta(\gamma) = h(g_\beta|\gamma)$, for all $\gamma < \beta^+$. But

$$(g_\alpha|\beta^+)(\gamma) = g_\alpha(\gamma)$$
$$= h(g_\alpha|\gamma)$$
$$= h((g_\alpha|\beta^+)|\gamma),$$

and consequently, by the uniqueness of g_β,

$$g_\beta = g_\alpha|\beta^+.$$

In particular, then, for all $\beta < \alpha$,

$$g_\beta(\beta) = (g_\alpha|\beta^+)(\beta) = g_\alpha(\beta).$$

Thus the rule which associates each α with the set X_α is as required by Corollary 6.19.

Proof (of Theorem 6.18)

First we prove uniqueness, on the assumption that some such function exists. Suppose that $g : \alpha \to X$ and $g' : \alpha \to X$ satisfy

$$g(\gamma) = h(g|\gamma) \quad \text{and} \quad g'(\gamma) = h(g'|\gamma).$$

for all ordinals $\gamma < \alpha$. We shall show that $g = g'$, by transfinite induction. Let

$$T = \{\gamma : \gamma < \alpha \quad \text{and} \quad g(\gamma) = g'(\gamma)\}.$$

T is a subset of α, which is a well-ordered set. We shall apply Corollary 6.16 to show that $T = \alpha$. First, T is not empty, because $g(0) = g'(0) = h(\emptyset)$. Next, suppose that $\gamma < \alpha$ and $\gamma \subseteq T$. Then $g(\delta) = g'(\delta)$ for each $\delta < \gamma$, i.e. $g|\gamma = g'|\gamma$. It follows from our assumption, then, that $g(\gamma) = h(g|\gamma) = h(g'|\gamma) = g'(\gamma)$, and so $\gamma \in T$. By Corollary 6.16, then, $T = \alpha$. Consequently $g(\gamma) = g'(\gamma)$ for every $\gamma < \alpha$, so that $g = g'$.

Now we demonstrate existence of the function g. Suppose that such a function (for given α and h) does not exist. Let B be the set of ordinals $\beta \leq \alpha$ such that there does not exist any function $g : \beta \to X$, satisfying $g(\gamma) = h(g|\gamma)$ for all ordinals $\gamma < \beta$. There is a least element of B, say β_0. Then for each ordinal $\delta < \beta_0$ there exists a function $g_\delta : \delta \to X$ such that

$$g_\delta(\gamma) = h(g_\delta|\gamma), \quad \text{for every ordinal } \gamma < \delta.$$

Indeed, we know that there is precisely one such function g_δ for each δ, by the proof of uniqueness, given above.

Now $\beta_0 \neq 0$, trivially, because the empty function in this case has the requisite properties.

Also β_0 cannot be a successor ordinal. To see this, suppose by way of contradiction that $\beta_0 = \delta^+$. There is a function $g_\delta : \delta \to X$ such that

$$g_\delta(\gamma) = h(g_\delta|\gamma), \quad \text{for every ordinal } \gamma < \delta.$$

Then g_δ is a transfinite sequence of length δ and, consequently, lies in the domain of h. We can therefore extend g_δ to a function $g : \delta^+ \to X$ by the inclusion of the ordered pair $(\delta, h(g_\delta))$. Then we have

$$g(\gamma) = g_\delta(\gamma) = h(g_\delta|\gamma) = h(g|\gamma), \quad \text{if } \gamma < \delta,$$

and

$$g(\delta) = h(g_\delta) = h(g|\delta).$$

Together, these assert:

$$g(\gamma) = h(g|\gamma), \quad \text{for every ordinal } \gamma < \beta_0,$$

and the existence of such a function g contradicts our supposition about β_0.

The last possibility is that β_0 may be a limit ordinal. We derive a contradiction in this case also, but it is a little more complicated this time. Consider the functions g_δ, for $\delta < \beta_0$. Suppose that $\delta < \xi < \beta_0$. Then g_ξ is a function from ξ to X, and $g_\xi|\delta$ is a function from δ to X satisfying, for every $\gamma < \delta$,

$$(g_\xi|\delta)(\gamma) = g_\xi(\gamma)$$

$$= h(g_\xi|\gamma)$$

$$= h((g_\xi|\delta)|\gamma).$$

By the uniqueness of g_δ, then, we have

$$g_\delta = g_\xi|\delta \quad \text{whenever } \delta < \xi < \beta_0.$$

Now we define $g : \beta_0 \to X$ by

$$g(\delta) = g_{\delta^+}(\delta), \quad \text{for all ordinals } \delta < \beta_0.$$

(Since β_0 is a limit ordinal, if $\delta < \beta_0$ then $\delta^+ < \beta_0$ also.) Then g extends each of the functions g_δ, and we have

$$g(\gamma) = h(g|\gamma), \quad \text{for all ordinals } \gamma < \beta_0,$$

since

$$g(\gamma) = g_{\gamma^+}(\gamma)$$
$$= h(g_{\gamma^+}|\gamma)$$
$$= h(g|\gamma).$$

Again, this contradicts our choice of β_0, and so we have finally contradicted our hypothesis that the function as described in the theorem does not exist.

▶ Definitions by transfinite induction, that is to say applications of the transfinite recursion theorem, are widespread in the foundations of mathematics, and are becoming increasingly used in mainstream mathematics also. In practice, the rule $\mathcal{R}$ (or function h) is given by cases, separately for zero, for successor ordinals and for limit ordinals. This was the case in Example 6.17, and the process used there of course has now been justified by Corollary 6.19. Let us now give another example.

Example 6.21

We have earlier hinted at an infinite process of construction of sets, starting with ω, and proceeding to $P(\omega)$, $P(P(\omega))$, and so on, each set being the power set of the preceding set. This can be generalised into a definition by transfinite induction which is very significant for foundational study.

Let

$$V_0 = \emptyset,$$

$$V_{\alpha^+} = P(V_\alpha), \quad \text{for each ordinal } \alpha,$$

and

$$V_\lambda = \bigcup\{V_\gamma : \gamma < \lambda\}, \quad \text{for each limit ordinal } \lambda.$$

These three clauses determine a rule $\mathcal{R}$:

with $\emptyset$ associate $\emptyset$,

with $\{X_\gamma : \gamma < \alpha^+\}$ associate $P(X_\alpha)$,

and

with $\{X_\gamma : \gamma < \lambda\}$ associate $\bigcup\{X_\gamma : \gamma < \lambda\}$,
where λ is a limit ordinal.

Thus, using Corollary 6.19, each ordinal α may be associated with a set V_α.

Let us investigate the sets V_α for small ordinals α:

$$V_0 = \emptyset,$$

$$V_1 = P(V_0) = P(\emptyset) = \{\emptyset\},$$

$$V_2 = P(V_1) = P(\{\emptyset\}) = \{\emptyset, \{\emptyset\}\},$$

$$V_3 = P(V_2) = P(\{\emptyset, \{\emptyset\}\}) = \{\emptyset, \{\emptyset\}, \{\{\emptyset\}\}, \{\emptyset, \{\emptyset\}\}\}.$$

Recalling our formal definition of natural numbers, we see that

$$V_0 = 0, \qquad V_1 = 1, \qquad V_2 = 2, \qquad V_3 = 3 \cup \{1\}, \quad \text{and so on.}$$

(The reader is recommended to work out V_4 for himself.) Suggested by the above is the general rule: $n \in V_{n^+}$, and this is easily proved (again an exercise). Consequently, $\omega \subseteq \bigcup \{V_n : n \in \omega\} = V_\omega$, so $\omega \in V_{\omega^+}$. Indeed, every subset of ω is contained in V_{ω^+}, and consequently $P(\omega) \in V_{\omega^{++}}$.

It is a consequence of the foundation axiom (ZF9) that given any set X there is an ordinal α such that $X \in V_\alpha$. Indeed, these two statements are equivalent. We shall not pursue this.

Example 6.22

We can use the transfinite recursion theorem to prove that the axiom of choice implies the well-ordering theorem (this has been proved by another, more cumbersome, method in the proof of Theorem 5.17). Recall that at .the beginning of Chapter 5 we described an intuitive procedure for picking out a countable subset from a given infinite set A: choose an element a_0 of A, then choose $a_1 \in A \backslash \{a_0\}$, then choose $a_2 \in A \backslash \{a_0, a_1\}$, etc. At that time we regarded this process as vague and unacceptable. We are now in a position, however, to give it respectability, and at the same time make it more general.

Suppose that A is any set. We assume (AC) and seek to describe a well-ordering of A. By (AC), the set $P(A) \backslash \{\emptyset\}$ has a choice function, f, say. Our transfinite construction will be by successive applications of f. It starts as follows:

$$g(0) = f(A), \qquad g(1) = f(A \backslash \{g(0)\}),$$

$$g(2) = f(A \backslash \{g(0), g(1)\}), \quad \text{etc.}$$

In general,

$$g(\gamma) = f(A \backslash \{g(\delta) : \delta < \gamma\}).$$

We are applying Theorem 6.18. This required an ordinal α acting as an upper bound. To obtain this, consider the set of all binary relations on A which well-order subsets of A. By Theorem 6.13 each such relation corresponds to a unique ordinal number, and, by means of the replacement axiom (ZF7), we see that there is a set of all ordinal numbers corresponding to well-orderings of subsets of A. Take α to be the smallest ordinal not in this set.

There is one more complication before we can obtain the result. Theorem 6.18 will yield a transfinite sequence of length α in this situation, except that we may run into trouble with the construction.

$$g(\gamma) = f(A \setminus \{g(\delta) : \delta < \gamma\}), \quad \text{for } \gamma < \alpha.$$

It may happen, for some $\gamma < \alpha$, that $A \setminus \{g(\delta) : \delta < \gamma\}$ is empty, and so $f(A \setminus \{g(\delta) : \delta < \gamma\})$ does not exist. We can avoid the problem by giving f an arbitrary value at $\emptyset$, say let $f(\emptyset) = b$, where $b \notin A$.

If it happens that $g(\gamma) = f(\emptyset) = b$, for some $\gamma < \alpha$, then it is not difficult to see that for all γ' with $\gamma < \gamma' < \alpha$, we have $g(\gamma') = b$ also. Also, it is easy to see that if $g(\gamma) \neq b$, then for $\delta < \gamma$, $g(\delta) \neq g(\gamma)$, since $g(\gamma) \notin \{g(\delta) : \delta < \gamma\}$.

There are two cases to consider. First, $g(\gamma) \neq b$ for all $\gamma < \alpha$. And second, $g(\gamma) = b$ for some $\gamma < \alpha$. In the first case the set $\{g(\gamma) : \gamma < \alpha\}$ is a subset of A, and is well-ordered through the bijection g between it and α, and so α is the ordinal number of a well-ordering of a subset of A. This contradicts the definition of α. Hence, the second case above must apply. Let β be the smallest ordinal such that $g(\beta) = b$. Then $\{g(\delta) : \delta < \beta\} = A$, and so A is well-ordered through the bijection g between it and the ordinal number β. The transfinite process builds, one step at a time, the bijection between A and an ordinal number. The well-ordering theorem thus follows from (AC).

▶ A similarly constructed proof can be given for the theorem (also proved earlier, as part of Theorem 5.17) that the well-ordering theorem implies Zorn's Lemma. See the exercises at the end of this section.

We give one example of a transfinite construction in the theory of groups, to illustrate how conveniently the ideas can fit into abstract algebra.

Example 6.23

Let G be an infinite group. We construct the *derived series* of G as follows. For any group H, the derived group H' is the subgroup of H generated by the set of all commutators in H (i.e. elements of the form $x^{-1}y^{-1}xy$ with $x, y \in H$). Let

$$G_0 = G,$$

$$G_{\gamma^+} = (G_\gamma)', \quad \text{for each ordinal } \gamma,$$

and

$$G_\lambda = \bigcap\{G_\gamma : \gamma < \lambda\}, \quad \text{for each limit ordinal } \lambda.$$

If G is abelian, the derived series is trivial, since $G_1 = \{1\}$. If G is not abelian, the derived series may be infinite, and it will continue until $G_{\gamma^+} = G_\gamma$ for some ordinal γ, after which it is constant. This could happen through G_γ being trivial, or through $G'_\gamma = G_\gamma \neq \{1\}$. A group G is said to be *soluble* if $G_\gamma = \{1\}$ for some *finite* ordinal γ. It can be easily proved by transfinite induction (with a little knowledge of group theory) that every G_γ is a normal subgroup of G.

▶ Our principal application of transfinite recursion is in defining the arithmetic operations on ordinals. In Chapter 4 we used the recursion theorem (Theorem 4.15) to define addition and multiplication of natural numbers. Here we generalise the process. We apply Corollary 6.19, since we wish to define operations on all ordinals.

Definitions
Addition. Fix an ordinal α, and define $s_\alpha(\beta)$ for all ordinals β by:

$$s_\alpha(0) = \alpha,$$

$$s_\alpha(\gamma^+) = (s_\alpha(\gamma))^+, \quad \text{for all ordinals } \gamma,$$

and

$$s_\alpha(\lambda) = \bigcup\{s_\alpha(\gamma) : \gamma < \lambda\}, \quad \text{for all limit ordinals } \lambda.$$

We write $\alpha + \beta$ for $s_\alpha(\beta)$.

Multiplication. Fix an ordinal α, and define $p_\alpha(\beta)$ for all ordinals β by:

$$p_\alpha(0) = 0,$$

$$p_\alpha(\gamma^+) = p_\alpha(\gamma) + \alpha, \quad \text{for all ordinals } \gamma,$$

and

$$p_\alpha(\lambda) = \bigcup\{p_\alpha(\gamma) : \gamma < \lambda\}, \quad \text{for all limit ordinals } \lambda.$$

We write $\alpha \cdot \beta$ or $\alpha\beta$ for $p_\alpha(\beta)$.

These operations have some familiar expected properties, and some awkward properties also.

Example 6.24

$1 + \omega = \omega$, and $\omega + 1 = \omega^+$. For the first, note that ω is a limit ordinal, so

$$1 + \omega = s_1(\omega) = \bigcup\{s_1(\gamma) : \gamma < \omega\}$$
$$= \bigcup\{s_1(n) : n \in \omega\}.$$

Now $s_1(n) = n^+$ (proof by induction on n – by methods of Chapter 4), and

$$\bigcup\{n^+ : n \in \omega\} = \bigcup\{n : n \in \omega\}$$
$$= \bigcup \omega$$
$$= \omega.$$

For the second,

$$\omega + 1 = s_\omega(1) = s_\omega(0^+) = (s_\omega(0))^+ = \omega^+.$$

Theorem 6.25
Ordinal addition is not commutative.

Proof
In the above example.

Theorem 6.26
Ordinal multiplication is not commutative.

Proof
$2\omega = \omega$, and $\omega 2 = \omega + \omega \neq \omega$. This is similar to the above example.

$$2\omega = p_2(\omega) = \bigcup\{p_2(n) : n \in \omega\}$$
$$= \bigcup\{2n : n \in \omega\}$$
$$= \bigcup\{n : n \in \omega\} = \bigcup \omega = \omega.$$

(Exercise: $p_2(n) = 2n$, for all $n \in \omega$, where the multiplication on the right-hand side is as defined in Section 4.3.) Also,

$$\omega 2 = p_\omega(2)$$
$$= p_\omega((0^+)^+)$$
$$= p_\omega(0^+) + \omega$$
$$= (p_\omega(0) + \omega) + \omega$$
$$= (0 + \omega) + \omega$$
$$= \omega + \omega$$
$$= s_\omega(\omega)$$
$$= \bigcup \{\omega + n : n \in \omega\}.$$

This last set is not equal to ω since, for example, $\omega^+ \subseteq \bigcup\{\omega + n : n \in \omega\}$ and certainly $\omega^+ \nsubseteq \omega$.

▶ The ideas used in these proofs generalise.

Theorem 6.27
(i) $\alpha + 1 = \alpha^+$, for every ordinal α.
(ii) $\alpha 2 = \alpha + \alpha$, for every ordinal α.

Proof
(i) $\alpha + 1 = s_\alpha(1) = s_\alpha(0^+) = s_\alpha(0)^+ = \alpha^+$.
(ii) Left as an exercise.

▶ Observe that in the above properties we have confirmation that there is sense in our previous suggestive notation, namely, $\omega + 1$, $\omega + 2$, ..., $\omega 2$, $\omega 3$, ... used in Section 6.1. The definitions we have now given agree exactly with those notations. Notice also that the definitions extend into the transfinite the definitions given in Chapter 4 of addition and multiplication on the set of abstract natural numbers, and so these new operations are identical to the previous ones in respect of elements of ω.

The arithmetic of ordinal numbers will not concern us in detail. Proofs of arithmetic properties of ordinals, such as the associative laws, cancellation laws and laws for inequalities, are generally by transfinite induction based on the recursive definitions of addition and multiplication. These proofs can be quite difficult, and can be lengthy. We give just one example of such a proof, and list some properties without proof.

Theorem 6.28
   (i) For every ordinal α, $1 \cdot \alpha = \alpha$.
   (ii) Addition and multiplication are associative.
   (iii) For any ordinals α, β, γ, $\alpha(\beta + \gamma) = \alpha\beta + \alpha\gamma$.
(The other distributive law does not hold; $(\alpha + \beta)\gamma$ is not in general equal to $\alpha\gamma + \beta\gamma$. Try to demonstrate this by finding a counterexample.)

Proof
   (i) Let α be some fixed ordinal. We apply Corollary 6.16 to show that $1 \cdot \beta = \beta$ for every $\beta \in \alpha$. Let $A = \{\beta \in \alpha : 1 \cdot \beta = \beta\}$. $A \neq \emptyset$, since $0 \in A$. We must show that, for each $\gamma \in \alpha$, if $\gamma \subseteq A$ then $\gamma \in A$. So let $\gamma \in \alpha$, and suppose that $\gamma \subseteq A$.
   Case 1: γ is a successor ordinal, say $\gamma = \delta^+$. Then $\delta < \gamma$, so $\delta \in \gamma$, and $\delta \in A$, since $\gamma \subseteq A$. Thus

$$1 \cdot \gamma = p_1(\delta^+) = p_1(\delta) + 1 = p_1(\delta)^+ = (1 \cdot \delta)^+ = \delta^+ = \gamma.$$

   Case 2: γ is a limit ordinal. Then

$$1 \cdot \gamma = p_1(\gamma) = \bigcup\{p_1(\delta) : \delta < \gamma\}$$
$$= \bigcup\{1 \cdot \delta : \delta < \gamma\}$$
$$= \bigcup\{\delta : \delta < \gamma\} = \gamma,$$

using the supposition that $\gamma \subseteq A$, so that $1 \cdot \delta = \delta$ for each $\delta < \gamma$.
   Corollary 6.16 now yields $A = \alpha$, i.e. $1 \cdot \beta = \beta$ for every $\beta \in \alpha$. But this is true for every ordinal α. Hence, $1 \cdot \beta = \beta$ holds for every ordinal β.

▶ As in ordinary arithmetic, addition consists in adding 1 an appropriate number of times ($m + n = m + 1 + 1 + \cdots + 1$), multiplication consists in adding the same number repeatedly ($mn = m + m + \cdots + m$). These are the sources of the recursive definitions. In ordinary arithmetic, there is a next stage. Repeated multiplication gives rise to exponentiation: $m^n = mmm \cdots m$. This works for ordinals too.

Definition
   Exponentiation. Fix an ordinal $\alpha \neq 0$, and define $e_\alpha(\beta)$ for all ordinals β by:

$$e_\alpha(0) = 1,$$

$$e_\alpha(\gamma^+) = e_\alpha(\gamma) \cdot \alpha, \quad \text{for all ordinals } \gamma,$$

and

$$e_\alpha(\gamma) = \bigcup \{e_\alpha(\gamma) : \gamma < \lambda\}, \quad \text{for all limit ordinals } \lambda.$$

This is justified by Corollary 6.19. We write α^β for $e_\alpha(\beta)$.

Examples 6.29

(a) $\omega^2 = e_\omega(1^+) = e_\omega(1) \cdot \omega = e_\omega(0^+) \cdot \omega$

$\qquad = (e_\omega(0) \cdot \omega) \cdot \omega = (1 \cdot \omega) \cdot \omega = \omega \cdot \omega.$

In general, for any ordinal α, $\alpha^2 = \alpha \cdot \alpha$.

(b) Inductively, $\alpha^n = \alpha \cdot \alpha \cdot \cdots \cdot \alpha$ (n factors), for any ordinal α and any finite ordinal n.

(c) $\omega^\omega = e_\omega(\omega)$

$\qquad = \bigcup \{e_\omega(n) : n \in \omega\}$

$\qquad = \bigcup \{\omega^n : n \in \omega\}.$

Thus ω^ω is the first ordinal larger than each of the ω^n for $n \in \omega$.

(d) The following standard laws for indices hold:

$$\alpha^{\beta+\gamma} = \alpha^\beta \cdot \alpha^\gamma,$$

and

$$(\alpha^\beta)^\gamma = \alpha^{\beta \cdot \gamma},$$

for any ordinals α, β, γ, with $\alpha \neq 0$. These require substantial inductive proofs, which we shall not go into.

(e) For every ordinal α, $1^\alpha = 1$. This is proved by an argument analogous to the proof of Theorem 6.28(i), and it is left as an exercise.

▶ A word of warning is in order here. We have previously considered another exponentiation operation, in Chapter 2. It should be clearly understood that this is something different. To emphasise the point, let us illustrate it by an example. ω^ω is equal to $\bigcup \{\omega^n : n \in \omega\}$, a union of countably many sets. Further, each ω^n is a countable set (see Exercise 6 at the end of this section). Thus ω^ω is a countable set. Compare this with the result from Chapter 2 that $\mathbb{N}^\mathbb{N}$ is uncountable. A careful distinction has to be made between *ordinal* exponentiation, which we have just defined, and exponentiation of sets or cardinal numbers as in Chapter 2. We shall mention this again when we discuss operations on cardinal numbers in the next section.

Exercises

1. Prove the following Corollaries of Theorem 6.15.
    (i) Let α be a non-zero ordinal and let A be a non-empty subset of α satisfying:

    if $\gamma \subseteq A$, then $\gamma \in A$,

    for each $\gamma < \alpha$. Then $A = \alpha$.
    (ii) Let α be a non-zero ordinal and let $\beta < \alpha$. Let $X = \{\gamma : \beta \leq \gamma < \alpha\}$, and let A be a subset of X such that $\beta \in A$ and satisfying: $\gamma \in A$ whenever $\delta \in A$, for all δ with $\beta \leq \delta < \gamma$. Then $A = X$.
2. In Example 6.23, prove that each G_γ is a normal subgroup of G. (This requires some knowledge of group theory.)
3. Prove the following, for all ordinals α.
    (i) $0 + \alpha = \alpha$.
    (ii) $1^\alpha = 1$.
    (iii) $\alpha 2 = \alpha + \alpha$.
4. Prove the following.
    (i) If $\alpha < \beta$, then $\alpha^+ < \beta^+$, for any ordinals α, β. (Use methods of Section 6.1.)
    (ii) If $\beta > 0$, then $\alpha < \alpha + \beta$, for any ordinals α, β. (Use transfinite induction on β.)
    (iii) If $\beta < \gamma$, then $\alpha + \beta < \alpha + \gamma$, for any ordinals α, β, γ. (Use transfinite induction on β.)
    (iv) Let X be a set of ordinals. Then $\bigcup X$ is an ordinal (Theorem 6.8) and $\alpha + \bigcup X = \bigcup \{\alpha + \beta : \beta \in X\}$, for any ordinal α.
    (v) $\alpha + (\beta + \gamma) = (\alpha + \beta) + \gamma$, for any ordinals α, β, γ.
5. Prove that $\alpha(\beta + \gamma) = \alpha\beta + \alpha\gamma$, for any ordinals α, β, γ. Find a counter-example which shows that $(\alpha + \beta)\gamma \neq \alpha\gamma + \beta\gamma$ in general.
6. Prove that ω^n is a countable set, for each $n \in \omega$. More generally, prove that if α and β are countable ordinals, then α^β is countable.
7. Fill in the details of the following proof that the well-ordering theorem implies Zorn's lemma. Let $(P, \leq)$ be a partially ordered (non-empty) set in which every chain has an upper bound. Suppose that P can be well-ordered, so that there is an ordinal α with a bijection $p : \alpha \to P$. Fix $t \notin P$. Define a transfinite sequence as follows. Let

$$c_0 = p(0),$$

and

$$\text{for } \beta > 0, \ c_\beta = \begin{cases} p(\gamma), & \text{where } \gamma \text{ is the smallest ordinal} \\ & \text{such that } c_\delta < p(\gamma) \text{ (in } P) \text{ for all} \\ & \delta < \beta, \text{ if such exists, and} \\ t & \text{otherwise.} \end{cases}$$

Show that
(a) There is an ordinal β such that $c_\beta = t$. Denote by η the least such.
(b) If $\beta < \eta$ then $\{c_\delta : \delta < \beta\}$ is a chain in P.
(c) η is a successor ordinal, and if $\eta = \xi^+$, say, then c_ξ is a maximal element of P.

6.3 Cardinal numbers

The difficulty about cardinal numbers, as we saw in Chapter 2, arises when we try to say what they are. Their basic properties are clear: with each set is associated a unique cardinal number, and sets which are equinumerous are associated with the same cardinal number. There is a parallel here between this situation and our earlier treatment of natural numbers. We seek to provide definitions based on the notions of standard set theory (and on nothing else) in such a way that the objects defined have the properties which we require. The properties of cardinal numbers generally are not as clear intuitively as those of natural numbers, so a formal definition of cardinal number is perhaps easier to accept.

There are two approaches to this matter. As we have noted, we could consider cardinal numbers to be 'equivalence classes' under the relation of equinumerosity, but there are objections to this, since such equivalence classes cannot be sets in ZF and are proper classes in VNB. However, this approach can be refined and a sensible definition made of cardinal numbers as equivalence classes of a sort. Axiom (ZF9), the foundation axiom, plays a vital role in this, through the transfinite sequence of the sets V_α described in Example 6.21. Details of this may be found in the book by Enderton.

We shall adopt the other approach, which is to use the ordinal numbers in a more direct way. Rather than taking the equivalence class itself to be the cardinal number of each set in the class, we shall choose one particular set from each equivalence class. How do we specify which one? We use the ordinal numbers. First of all let us pick out some ordinals with a particular property.

Definition

An infinite ordinal number α is an *aleph* (or an *initial ordinal*) if α is not equinumerous with any smaller ordinal.

Examples 6.30

(a) ω is an aleph, obviously. It is the smallest aleph, and the notation $\aleph_0$ now becomes clear.

(b) What is the next aleph after ω? It cannot be equinumerous with ω, so it must be uncountable. Thus ω_1 (the least uncountable ordinal) is the next aleph. This is sometimes denoted by $\aleph_1$. Notice that there are many ordinals between $\aleph_0$ and $\aleph_1$.

   Is there an unending sequence of alephs? The answer is in the affirmative, in a very strong sense, namely, that there is a transfinite sequence of alephs, indexed by the ordinals themselves. This is an easy transfinite construction based on the following theorem.

Theorem 6.31
Given any ordinal α, there is an aleph greater than α.

Proof
This generalises an argument given at the end of Section 6.1 (which dealt with the case where $\alpha = \omega$). The result is trivial if α is finite, so suppose that α is infinite. We first prove that there is a set of all ordinals γ such that $\gamma \leqslant \alpha$. We apply the replacement axiom (ZF7) using the formula $\mathscr{F}(R, \gamma)$:

'R is a relation which well-orders a subset A of α,

γ is an ordinal and $\gamma = \mathrm{ord}\,(A, R)$.'

This can be translated into the precise form required for (ZF7), and it certainly determines a function. By (ZF6), there is a set

$$W = \{R \in P(\alpha \times \alpha) : R \text{ is a well-ordering of a subset of } \alpha\}.$$

The image of the set W under the function determined by the formula $\mathscr{F}(R, \gamma)$ is the required set of all ordinals γ such that $\gamma \leqslant \beta$. Demonstration of this is left as an exercise.

   Let $D = \{\gamma : \gamma \leqslant \alpha\}$. By Theorem 6.8, since D is a set of ordinals, $\bigcup D$ is an ordinal, and $(\bigcup D)^+$ is certainly larger than every ordinal in D. Denote $(\bigcup D)^+$ by α_0. Then $\alpha_0 \notin D$, so α_0 is not equinumerous with α. Hence, the set $\{\gamma \in \alpha_0^+ : \alpha < \gamma \text{ and } \gamma \text{ is not equinumerous with } \alpha\}$ is not empty. It has a least element, therefore, which must be an aleph and must be larger than α.

Corollary 6.32
Given any ordinal α, there is a sequence $\{\aleph_\gamma : \gamma < \alpha\}$ of alephs such that for $\gamma < \delta < \alpha$, $\aleph_\gamma < \aleph_\delta$.

Proof

We apply Theorem 6.18 to the following construction. Let

$$\aleph_0 = \omega,$$

$\aleph_{\gamma^+} =$ the smallest aleph greater than $\aleph_\gamma$, for each ordinal γ,

and

$$\aleph_\lambda = \bigcup\{\aleph_\gamma : \gamma < \lambda\}, \text{ for each limit ordinal } \lambda.$$

Verification that $\bigcup\{\aleph_\gamma : \gamma < \lambda\}$ is in fact an aleph and is distinct from all the $\aleph_\gamma$ (for $\gamma < \lambda$), is left as an exercise.

Corollary 6.33

To each ordinal α may be associated an aleph, denoted by $\aleph_\alpha$, such that for $\gamma < \delta$ we have $\aleph_\gamma < \aleph_\delta$. Moreover, every aleph is one of the $\aleph_\alpha$ for some ordinal α. Notice the distinction between α (an arbitrary ordinal) and $\aleph_\alpha$ (an aleph) which is normally larger than α.

▶ We have gone to some trouble over alephs because these are to be the 'special' sets which will be our cardinal numbers. Let X be any set. The axiom of choice (through the well-ordering theorem) implies that there is a relation R such that (X, R) is well-ordered. By Theorem 6.13 there is a unique ordinal number which is order isomorphic to (X, R). For our purposes here, however, we need only the information that there is an ordinal number which is equinumerous with X. The least such ordinal number will be an aleph (in the case where X is infinite), will be uniquely determined by X, and will be the cardinal number of X.

Definition

Given any set X, *the cardinal number of X*, written card X, is the smallest ordinal number which is equinumerous with X.

▶ Notice that the preceding discussion yields the basic property: every set X has a uniquely determined cardinal number. At the same time it should be made clear that the axiom of choice was used in that discussion. We did mention earlier an alternative method for defining cardinal numbers which depends on the foundation axiom and not on the axiom of choice. However, the properties of cardinal numbers turn out to be significantly bound up with the axiom of choice. We have already seen some evidence of this in earlier chapters. For example, (AC) is necessary for the result that, for any sets X and Y, either $X \leqslant Y$ or $Y \leqslant X$

(Theorem 5.28(v)), which has consequences for the ordering of cardinal numbers (see Theorem 6.36). Also recall Theorem 2.38, that for any infinite cardinal numbers κ and λ, $\kappa + \lambda = \kappa\lambda = \max(\kappa, \lambda)$ which depended on (AC). It certainly is possible to develop a theory of cardinal numbers independent of the axiom of choice in which these theorems are not derivable. We shall be more conventional, however, and for the remainder of this chapter we shall *assume that the axiom of choice holds*. Consequently, every set has a cardinal number under our definition, and we shall be able to develop the arithmetic of cardinal numbers to match what we found in Chapter 2.

Our definition makes no reference to alephs, but it is easy to see that if X is an infinite set then card X is an aleph. If X is a finite set then card X is clearly a finite ordinal.

Definition
We define the term *cardinal number* as follows. All finite ordinals are cardinal numbers, all alephs are cardinal numbers and there are no other cardinal numbers. (As with ordinals, we abbreviate 'cardinal number' to 'cardinal'.)

The use of lowercase Greek letters to denote cardinals will obviously be consistent, and we shall follow our earlier practice (in Chapter 2) and use the letters κ, λ, μ, ν for cardinals.

► Now let us derive some properties of cardinals.

Theorem 6.34
Every set of cardinals is well-ordered by $\subseteq$.

Proof
Every set of cardinals is a set of ordinals, so this is an immediate consequence of Theorem 6.8(i).

Notation
As with ordinals we use the symbols $\leqslant$ and $<$ to denote the order, rather than $\subseteq$, in order to emphasise that this is an order by magnitude.

Theorem 6.35
Given any set X of cardinals, $\bigcup X$ is a cardinal and is the smallest cardinal greater than or equal to every member of X.

Proof

If X contains a greatest element, κ, say, then $\bigcup X = \kappa$ (see Corollary 6.9). The result in this case is now immediate. If X does not contain a greatest element, then, by Corollary 6.9, $\alpha = \bigcup X$ is the smallest *ordinal* greater than or equal to every member of X. We must show that α is a cardinal number. First, α cannot be finite, since X has no greatest element. Suppose that α is not an aleph, i.e. that $\alpha \sim \beta$ for some ordinal $\beta < \alpha$. By the choice of α, there must exist a cardinal $\mu \in X$ with $\beta \leqslant \mu$. Since X has no greatest element, there is $\nu \in X$ with $\mu < \nu$ (so that $\mu \not\sim \nu$, since ν is a cardinal). Hence we have $\alpha \leqslant \beta \leqslant \mu \leqslant \nu \leqslant \alpha$, which by the Schröder–Bernstein theorem (Theorem 2.18) yields the conclusion $\mu \sim \nu$, which is a contradiction. Consequently, α is an aleph, so $\bigcup X$ is a cardinal number, and the proof is complete.

Theorem 6.36
For any sets X and Y:
(i) Card X = card Y if and only if $X \sim Y$.
(ii) Card $X \leqslant$ card Y if and only if $X \leqslant Y$.

Proof
(i) $X \sim$ card X and $Y \sim$ card Y, so card X = card Y implies $X \sim Y$. (See Theorem 2.1.) Conversely, let $X \sim Y$. For any ordinal α, $\alpha \sim X$ if and only if $\alpha \sim Y$, so the smallest α with $\alpha \sim X$ must also be the smallest with $\alpha \sim Y$. Hence, card X = card Y.

(ii) Suppose that card $X \leqslant$ card Y. Then card X is a subset of card Y, so we may define an injection $X \to Y$ by $X \to$ card $X \to$ card $Y \to Y$, where the first and third arrows stand for bijections and the second for set inclusion. Thus $X \leqslant Y$. Conversely, suppose that there is an injection $f : X \to Y$. Let $A = f(X)$, so that card A = card X. There exists a well-ordering R of Y such that ord (Y, R) = card Y. Now $A \subseteq Y$, so ord $(A, R) \leqslant$ ord (Y, R) (see Exercise 11 on page 205). Hence,

card X = card $A \leqslant$ ord $(A, R) \leqslant$ ord (Y, R) = card Y,

and so card $X \leqslant$ card Y, as required.
Notice that (ii) implies (i) by virtue of the Schröder–Bernstein theorem.

▶ Now we come to a point where there may be confusion. We saw in Chapter 2 that there is an arithmetic of cardinal numbers. We defined there addition, multiplication and exponentiation of cardinals. Now that we have defined cardinals as ordinals of a particular kind we must check that those definitions are still applicable, and we must investigate whether they bear any relation to the definitions of the corresponding operations on ordinals, introduced in Section 6.2.

Definition

Let κ and λ be any two cardinal numbers.
(i) $\kappa + \lambda$ is card $(A \cup B)$, where A and B are any sets with card $A = \kappa$, card $B = \lambda$ and $A \cap B = \emptyset$.
(ii) $\kappa \cdot \lambda$ is card $(A \times B)$, where A and B are any sets with card $A = \kappa$ and card $B = \lambda$.
(iii) κ^{λ} is card (A^{B}), where A and B are any sets with card $A = \kappa$ and card $B = \lambda$.

▶ These definitions are the same as in Chapter 2, and it is easy to see that they are not at all problematic in the present context. But a strong warning should be noted that these are *not* the same operations as the operations of ordinal arithmetic introduced in Section 6.2, even though the cardinal numbers κ and λ are in fact ordinals.

Examples 6.37

(a) Under *cardinal* addition, as above:

$$\omega + \omega = \omega.$$

We can see this by the methods of Chapter 2, where we proved that $\aleph_0 + \aleph_0 = \aleph_0$.

Under *ordinal* addition, however:

$$\omega + \omega = \omega 2 \neq \omega.$$

(b) Likewise for multiplication:

$$\omega \cdot \omega = \omega \text{ as cardinal multiplication}$$

and

$$\omega \cdot \omega = \omega^2 \neq \omega \text{ as ordinal multiplication.}$$

(c) For exponentiation the situation is slightly different: $\omega^{\omega} = \aleph_0^{\aleph_0}$ by cardinal exponentiation is uncountable, whereas ω^{ω} (ordinal exponentiation) is a *countable* ordinal.

▶ We shall not develop the arithmetic of cardinals, but let us recall another result from an earlier chapter.

Theorem 6.38

For any infinite cardinal numbers κ and λ, if $\kappa \leq \lambda$ then $\kappa + \lambda = \lambda$ and $\kappa \cdot \lambda = \lambda$.

Proof
See Theorem 2.38 and Corollary 5.16. Of course the proof is dependent on the axiom of choice.

▶ As we already know (from Chapter 2), there is some uncertainty about how exponentiation fits into the hierarchy of cardinals. In Chapter 2 we mentioned the continuum hypothesis:

(CH) Every subset of $\mathbb{R}$ is either countable or equinumerous with $\mathbb{R}$.

In terms of cardinal numbers this may be restated as follows.

(CH*) There is no cardinal number lying strictly between $\aleph_0$ and $2^{\aleph_0}$.

Further, in view of what we know about the ordering of cardinal numbers, this is equivalent to:

(CH**) $2^{\aleph_0} = \aleph_1$.

Of course, $\aleph_1$ is by definition the smallest cardinal larger than $\aleph_0$.

This hypothesis arose out of the knowledge that $P(\mathbb{N})$ has a strictly larger cardinal number than $\mathbb{N}$, and out of the inability of mathematicians to find any sets with cardinal numbers in between. These ideas can be generalised. We know (Theorem 2.16) that for any set A, the power set of A has a strictly larger cardinal number than A. It has been conjectured that there is no cardinal number strictly between card A and card $P(A)$, for all infinite sets A. (Notice that this fails in general for finite sets A.) This conjecture is called the *generalised continuum hypothesis*:

(GCH) Given any infinite set A and any set B, if $A \leq B \leq P(A)$ then either $B \sim A$ or $B \sim P(A)$.

Perhaps a more familiar version of (GCH) is the following

(GCH*) Given any cardinal number κ, there is no cardinal number lying strictly between κ and 2^κ.

We have not stated this as our standard version, since it presupposes the axiom of choice, according to our procedure for defining cardinal numbers. Avoiding this complication (as in the statement of (GCH)) makes consideration of the logical relationship between (AC) and (GCH) rather easier. Another version, called Cantor's hypothesis on alephs, perhaps makes this clearer.

(GCH**) For every ordinal α, $\aleph_{\alpha+1} = 2^{\aleph_\alpha}$.

In the presence of (AC), every infinite set has a cardinal number which is an aleph, and all these three above are equivalent. Indeed, we can state the position more precisely.

Theorem 6.39
In ZF, (GCH) is equivalent to the conjunction of (GCH**) and (AC). In particular, (GCH) implies (AC).

Proof
Omitted. See the book by Sierpinski (Section xvi.5).

▶ (GCH) is very neat and tidy. It would be very nice if it were true. But is it true? Our intuition is certainly inadequate to answer this immediately. It is not clearly true or clearly false. Thus we seek to demonstrate its truth or falsity on the basis of other principles. This has been one of the major problems in mathematics since the end of the nineteenth century. The methods of formal axiomatic set theory and models have been brought to bear on it, just as they have been applied to the question of the truth of the axiom of choice, with similar conclusions.

Theorem 6.40
(i) (Gödel 1938) Given that ZF is consistent, the system obtained by adding (GCH) as an additional axiom is also consistent.
(ii) (Cohen 1963) Given that ZF is consistent, (GCH) cannot be derived as a consequence of the ZF axioms.
(iii) (Cohen 1963) Given that ZF is consistent, (GCH) cannot be derived as a consequence of (AC) together with the ZF axioms.

The proofs of these, like the corresponding ones for results about the axiom of choice, are very difficult, involving constructions of models for ZF.

▶ As with (AC), the axioms of ZF just provide no guide as to the truth of (GCH). Consequently we are thrown back on intuition, which we have already noted as inadequate. The current position amongst mathematicians is one of scepticism about (GCH), while searching for other principles which may be clearer intuitively, on the basis of which (GCH) may be proved or disproved.

The position regarding the particular case which we have called (CH) is no better. Since (GCH) is consistent with ZF, so is (CH). Cohen's methods also yield the independence of (CH) from the ZF axioms, and indeed from ZFC.

Lack of intuition as to the truth or falsity of (GCH) and (CH) is a substantial impediment to further study of infinite cardinals. We do not know where $2^{\aleph_0}$ fits in to the list $\aleph_0, \aleph_1, \ldots$, indexed by the ordinals. For the foundationist this does not matter greatly. It provides him with interesting work – indeed, he may establish one theory with (GCH) and an entirely separate theory without (GCH), in the knowledge that both are consistent (provided that ZF itself is, of course). With regard to the latter, it is of interest to note that it is consistent (with ZFC) to suppose that $2^{\aleph_0} = \aleph_\alpha$, where $\aleph_\alpha$ is any aleph which has no countable cofinal subset. (A subset A of an ordered set $(X, \leq)$ is *cofinal* if for each $x \in X$ there is $a \in A$ with $x \leq a$. See Exercise 20 on page 96). We shall return idea shortly, but let us just state just now that $2^{\aleph_0}$ can be, for example, any $\aleph_n$ for finite n, and cannot be $\aleph_\omega$.

On the other hand, for mathematicians in general it is very frustrating not to know whether (GCH) is true. Clearly, the set $\mathbb{R}$ is central to all mathematical analysis; hence (CH) is of concern. Further, function spaces, for example, are subsets of sets of the form X^Y, and are therefore subsets of sets with cardinal numbers of the form $\aleph_\alpha^{\aleph_\beta}$, say. Without (GCH) little can be said about such cardinals, and perhaps less about the cardinal numbers of the subsets. *With* (GCH) the situation here is clear.

Theorem 6.41
Let α and β be ordinals, with $\alpha \leq \beta$. Then $\aleph_\alpha^{\aleph_\beta} = 2^{\aleph_\beta} = \aleph_{\beta+1}$, on the assumption of (GCH).

Proof
By the construction of the sequence of alephs, since $\alpha \leq \beta$ we have $\aleph_\alpha \leq \aleph_\beta$, and so $2^{\aleph_\alpha} \leq 2^{\aleph_\beta}$ (Exercise 3 on page 72). We therefore have

$$2 < \aleph_\alpha < 2^{\aleph_\alpha} \leq 2^{\aleph_\beta}.$$

Now

$$2^{\aleph_\beta} \leqslant \aleph_\alpha^{\aleph_\beta} \qquad \text{(methods of Chapter 2 again),}$$

$$\leqslant (2^{\aleph_\beta})^{\aleph_\beta} \qquad \text{by the above (see Exercise 9 on page 81),}$$

$$= 2^{(\aleph_\beta^2)} \qquad \text{(Exercise 13 on page 81),}$$

$$= 2^{\aleph_\beta} \qquad \text{by Corollary 5.16.}$$

Hence, by the Schröder–Bernstein theorem, we obtain

$$\aleph_\alpha^{\aleph_\beta} = 2^{\aleph_\beta}, \quad \text{as required.}$$

For the latter equality we need only apply (GCH**) directly, to obtain $2^{\aleph_\beta} = \aleph_{\beta+1}$.

Corollary 6.42

Given (GCH), if X and Y are sets with $X \leqslant Y$ and A is a subset of X^Y, then either $A \leqslant Y$ or $A \sim X^Y$.

▶ Let us return to the sequence of alephs, and consider, in particular, $\aleph_\omega$. By the construction scheme, $\aleph_\omega = \bigcup\{\aleph_n : n \in \omega\}$. Although $\aleph_\omega$ is certainly uncountable, it has a countable cofinal subset. Actually finding such a subset should help to clarify the picture. Certainly, $\aleph_0, \aleph_1, \aleph_2, \ldots$ is an increasing sequence of ordinals, so in particular each $\aleph_n$ is a proper subset of $\aleph_{n+1}$ and hence, by Corollary 6.6, we have $\aleph_n \in \aleph_{n+1}$. Consequently, for each $n \in \omega$, $\aleph_n \in \bigcup\{\aleph_n : n \in \omega\}$, i.e. $\aleph_n \in \aleph_\omega$. We show that the set $A = \{\aleph_n : n \in \omega\}$ is cofinal in $\aleph_\omega$. Let $\alpha \in \aleph_\omega$. Then $\alpha \in \bigcup\{\aleph_n : n \in \omega\}$, so $\alpha \in \aleph_m$ for some $m \in \omega$. This means that $\alpha < \aleph_m$, and thus A is cofinal in $\aleph_\omega$. Clearly, A is countable. Because $\aleph_\omega$ is a countable union of smaller cardinals, we can pick out a countable subset which is cofinal. Recall in passing that it is this property which yields the impossibility of $2^{\aleph_0}$ being equal to $\aleph_\omega$.

But this property of $\aleph_\omega$ is a particular case of a more general characteristic that cardinals may or may not have.

Examples 6.43

(a) Consider $\aleph_1$. No subset of $\aleph_1$ which is cofinal can be countable. To see this, suppose that X is a countable cofinal subset of $\aleph_1$, and let $X = \{\alpha_1, \alpha_2, \ldots\}$ in increasing order. For each $i \in \mathbb{N}$, let $A_i = \{\gamma \in \aleph_1 : \gamma \leqslant \alpha_i\}$. Then, since X is cofinal in $\aleph_1$, we have

$$\aleph_1 = \bigcup\{A_i : i \in \mathbb{N}\}.$$

But each A_i is countable, since each α_i is a countable ordinal (being smaller than the first uncountable ordinal). Hence, $\aleph_1$ is a countable union of countable sets, and is itself countable. Here is the contradiction which yields the required conclusion.

(b) The above argument generalises to show that for each finite $n > 1$ $\aleph_n$ has no cofinal subset of cardinal number less than $\aleph_n$. This is left as an exercise, dependent on the theorem that if card $I < \aleph_n$ and card $X_i < \aleph_n$ for each $i \in I$, then card $(\bigcup\{X_i : i \in I\}) < \aleph_n$. This is because we must have $I \leqslant \aleph_{n-1}$ and $X_i \leqslant \aleph_{n-1}$, and since

$$\bigcup\{X_i : i \in I\} \leqslant \aleph_{n-1} \times I$$

$$\leqslant \aleph_{n-1} \times \aleph_{n-1},$$

we have, by Corollary 5.16,

$$\bigcup\{X_i : i \in I\} \leqslant \aleph_{n-1},$$

so that

$$\text{card } (\bigcup\{X_i : i \in I\}) \leqslant \aleph_{n-1} < \aleph_n.$$

Definition
A cardinal number κ is *singular* if κ contains a cofinal subset with cardinal number strictly less than κ. Otherwise, κ is *regular*.

Examples 6.44
(a) $\aleph_\omega$ is singular.
$\aleph_n$ is regular for each $n \in \omega$.
(b) $\aleph_\lambda$ is singular whenever λ is a limit ordinal, provided that $\lambda < \aleph_\lambda$. To see this we need only extend the argument we applied above to find a countable cofinal subset of $\aleph_\omega$, since $\aleph_\lambda = \bigcup\{\aleph_\gamma, \gamma < \lambda\}$. The possibility is open that $\lambda = \aleph_\lambda$ for some limit ordinals λ. In that case we can reach no general conclusion as to whether $\aleph_\lambda$ is singular or regular. Such an ordinal which is regular is said to be *weakly inaccessible*.
(c) $\aleph_{\alpha+1}$ is regular, for every ordinal α. This requires an argument similar to that in Example 6.43(b).

▶ A cardinal which is weakly inaccessible must be very large indeed. Let us try to give some idea of how large by considering the condition $\lambda = \aleph_\lambda$ which must be satisfied by such cardinals. We can construct a cardinal λ such that $\lambda = \aleph_\lambda$ as follows. By induction we may construct a sequence of cardinals $\{\kappa_n : n \in \omega\}$ by:

$$\kappa_0 = \aleph_0,$$

and

$$\kappa_{n+1} = \aleph_{\kappa_n}, \quad \text{for each } n \in \omega.$$

Now $\bigcup\{\kappa_n : n \in \omega\}$ is a cardinal number. Let us denote it by λ. Then $\aleph_\lambda = \lambda$, by the following argument. $\{\kappa_n : n \in \omega\}$ is a set of ordinals cofinal in λ, so $\{\aleph_{\kappa_n} : n \in \omega\}$ is a set of ordinals cofinal in $\{\aleph_\gamma : \gamma < \lambda\}$. Hence, by Exercise 8 on page 205, we have

$$\bigcup\{\aleph_{\kappa_n} : n \in \omega\} = \bigcup\{\aleph_\gamma : \gamma < \lambda\},$$

i.e.

$$\bigcup\{\kappa_{n+1} : n \in \omega\} = \aleph_\lambda,$$

i.e.

$$\lambda = \aleph_\lambda,$$

since $\kappa_0 \leqslant \kappa_1$ and $\lambda = \bigcup\{\kappa_n : n \in \omega\}$. We have therefore proved:

Theorem 6.45
There is an ordinal λ such that $\aleph_\lambda = \lambda$.

▶ Observe that the cardinal $\aleph_\lambda$ as constructed above is *not* weakly inaccessible, because it is not regular. The set $\{\kappa_n : n \in \omega\}$ is a cofinal subset with smaller cardinal number, so $\aleph_\lambda$ is singular. But we obtain some idea of the size that a weakly inaccessible cardinal must be when we think that this λ is, in effect, the limit of the sequence.

$$\aleph_0, \aleph_{\aleph_0}, \aleph_{\aleph_{\aleph_0}}, \dots.$$

Large cardinals are a subject of study in their own right, and in this book we can do no more than describe something of its beginnings. One reason for interest in them is that propositions are found which may help to elucidate set theory, in the following sense. (GCH) is known to be consistent and independent, but it is not intuitively clear, so we seek other principles which are consistent and independent and which may be intuitively clearer, in order to help with our understanding where there are gaps. As an example we can cite the proposition that there exists a weakly inaccessible cardinal. This is known to be independent of the axioms of ZF plus (AC). Thus this proposition is a candidate (at least) to be an additional axiom for set theory, and investigation of its consequences may shed some light on set theory itself. Intuitive justification for the existence of large cardinals is perhaps difficult, but there is an analogy with axiom (ZF8), the infinity axiom, in the sense that the existence of a particular set is postulated in order to make a jump from smaller sets to larger sets.

We have come across weakly inaccessible cardinals, which apparently must be very large. For the sake of completeness let us say what is meant by an inaccessible cardinal, as it makes rather more sense of the terminology.

Definition
A cardinal number κ is *inaccessible* if it is regular, it is greater than $\aleph_0$ and, for any cardinal $\mu < \kappa$, we have $2^\mu < \kappa$.

Notice that the last condition holds for $\aleph_0$. The first inaccessible cardinal (if there is one) is the next largest regular cardinal which shares this property.

It can be shown (though it is not easy) that every inaccessible cardinal is weakly inaccessible. It is not possible to prove in ZFC that inaccessible cardinals exist. One principal reason for interest in inaccessible cardinals is that they provide models for ZF set theory. Let us just state the theorem without comment.

Theorem 6.46
Provided that ZF is consistent, for any inaccessible cardinal κ the set V_κ (as described in Example 6.21) is a model for ZFC.

► Finally, let us briefly consider measurable cardinals, since they seem to have entered the general mathematical consciousness in recent years. The idea of a measure originated in the context of sets of real numbers, and we mentioned it in Chapter 5. There we described Lebesgue measure, which is a real valued function μ whose domain is a subset of $P(\mathbb{R})$, which is invariant under translations and which is countably additive, i.e.

$$\mu(\bigcup\{X_i : i \in \mathbb{N}\}) = \sum_{i \in \mathbb{N}} \mu(X_i),$$

where $\{X_i : i \in \mathbb{N}\}$ is any family of pairwise disjoint subsets of $\mathbb{R}$. In the context of cardinals, the idea of a measure has to be slightly different, for reasons which we shall not pursue.

Definition

A cardinal number κ is *measurable* if $\kappa > \aleph_0$ and there is a function $\mu : P(\kappa) \to \{0, 1\}$ such that $\mu(\kappa) = 1$, which is κ-additive, i.e. given any family $\{X_\gamma : \gamma < \lambda\}$ of pairwise disjoint subsets of κ indexed by a cardinal number $\lambda < \kappa$,

$$\mu \left(\bigcup \{X_\gamma : \gamma < \lambda\} \right) = \sum_{\gamma < \lambda} \mu(X_\gamma).$$

▶ Notice that the function which takes value **0** everywhere is trivially κ-additive so we stipulate that $\mu(\kappa) = 1$ in order to exclude this possibility.

It may help to clarify the meaning of κ-additivity to point out that, since the values of μ are **0** and **1**, the equation above is equivalent to saying that, in *any* family $\{X_\gamma : \gamma < \lambda\}$ as above, either $\mu(X_\gamma) = 0$ for all $\gamma < \lambda$ and $\mu(\bigcup \{X_\gamma : \gamma < \lambda\}) = 0$, or $\mu(X_\gamma) = 1$ for *precisely one* of the X_γ, and $\mu(\bigcup \{X_\gamma : \gamma < \lambda\}) = 1$.

Using the methods of Chapter 3, it is not difficult to show that there is a two-valued $\aleph_0$-additive non-trivial measure on $\aleph_0$ (i.e. a function $P(\aleph_0) \to \{0, 1\}$ as described above). If we choose a non-principal maximal ideal I in the Boolean algebra $(P(\mathbb{N}), \subseteq)$ and let $\mu(X) = 0$ if and only if $X \in I$ and $\mu(X) = 1$ otherwise (for $X \subseteq \mathbb{N}$), then this function μ does the job. Thus $\aleph_0$ would satisfy the definition of measurable cardinal if it were not specifically excluded.

It is not easy to see that measurable cardinals (which by definition are greater than $\aleph_0$) must be very large. But it is known that all measurable cardinals are inaccessible, for example. And a corollary of that result, of course, is that it cannot be proved in ZFC that there exist any measurable cardinals.

The results above generally depend on the axiom of choice, and do not depend on the generalised continuum hypothesis. It is perhaps of interest to note that, in the absence of the axiom of choice, the whole picture may be entirely different. In Chapter 5 we mentioned the axiom of determinateness and noted some of its consequences, particularly that it contradicts (AC). Another consequence of the axiom of determinateness is the surprising proposition that $\aleph_1$ is a measurable cardinal. In the presence of (AC), a measurable cardinal must be very large. Without it, and with (AD), there is a very small measurable cardinal. It was noted in Chapter 5 that (AD) is under suspicion and is not known to be consistent with ZF. It is also fair to say that the assertion that there is a measurable cardinal is under suspicion. There is little intuitive justification for it, and it is not known whether it is consistent with ZF.

Reality (with intuition as guide) has now almost completely disappeared. What are under discussion in this field are logical relationships between certain assertions, some of which are called axioms. It is small wonder that there is philosophical disputation about what is meant by existence in a mathematical sense. It cannot be doubted, however, that (to take but one example) the demonstration that (AD) implies that $\aleph_1$ is a measurable cardinal *is* mathematics and is of interest to the mathematician, even though it may be shown in the future that it is either trivial or meaningless. Although basic intuitions do not figure in this mathematics, it has to be accepted that its origins lie in simply-framed questions, like 'what is the nature of sets?' and that the mathematics does attempt to answer such questions. The fact that these questions are very difficult means that the mathematics which arises turns out to be very complicated. It also means that mathematicians are constantly shooting in the dark, in the hope of some time striking relevance and clarification.

Exercises

1. Let α and α_0 be infinite ordinals such that $\alpha < \alpha_0$ and α is not equinumerous with α_0. Prove that the set $\{\gamma \in \alpha_0 : \alpha < \gamma$ and $\alpha \not\sim \gamma\}$ has a least element which is an aleph.
2. Let λ be a limit ordinal. Prove that $\bigcup\{\aleph_\gamma : \gamma < \lambda\}$ is an aleph.
3. Prove the result in Corollary 6.33 that every aleph is one of the $\aleph_\alpha$ for some ordinal α.
4. Show that for every infinite set X, card X is an aleph.
5. Prove that for $n \in \omega$ $(n \neq 0)$, $\aleph_n$ has no countable cofinal subset. Prove further that if λ is a limit ordinal then $\aleph_\lambda$ has a cofinal subset A with card $A = $ card λ.
6. Assuming the generalised continuum hypothesis, what can be said about $\aleph_\alpha^{\aleph_\beta}$, when $\beta < \alpha$?
7. Prove the following.
   (i) $\aleph_{\alpha+1}$ is regular, for all ordinals α.
   (ii) If λ is a limit ordinal and $\lambda < \aleph_\lambda$, then $\aleph_\lambda$ is singular.
8. Prove that a cardinal number κ is measurable if and only if there is a function $\mu : P(\kappa) \to \{0, 1\}$ such that
   (a) $\mu(\kappa) = 1$,
   (b) $\mu(\{\gamma\}) = 0$, for every $\gamma \in \kappa$,
   (c) $\mu(\kappa \backslash X) = 1 - \mu(X)$, for every subset X of κ,
   and
   (d) $\mu(\bigcup\{X_\gamma : \gamma < \lambda\}) = 0$ if $\mu(X_\gamma) = 0$, for each $\gamma < \lambda$, where λ is any cardinal smaller than κ and $X_\gamma \subseteq \kappa$, for each ordinal $\gamma < \lambda$.

9. Given the existence of a non-principal maximal ideal I in the Boolean algebra $(P(\mathbb{N}), \subseteq)$, show in detail that there is a two-valued countably additive measure on $\aleph_0$.

Further reading

Barwise [1] See page 191.
Cohen & Hersh [6] See page 162.
Drake [8] An advanced book, requiring knowledge of mathematical logic, containing much about large cardinals.
Enderton [9] See page 162.
Fraenkel, Bar-Hillel & Levy [10] See page 162.
Halmos [12] See page 162.
Kuratowski & Mostowski [18] See page 107.
Sierpinski [22] See page 81.

HINTS AND SOLUTIONS TO SELECTED EXERCISES

Section 1.1 (p. 17)

2. Let A be the set $\{n \in \mathbb{N}: (x+y)+n = x+(y+n) \text{ for all } x, y \in \mathbb{N}\}$. Apply (P5) to A, using the property (A).

3. Similar to Exercise 2.

4. Apply (P5) to the set $\{0\} \cup \{n \in \mathbb{N}: n = m', \text{ for some } m \in \mathbb{N}\}$. Use (P3), (A) and (M).

5. For each $m \in \mathbb{N}$, show (using (P5)) that the set $\{n \in \mathbb{N}: m \leqslant n \text{ or } n \leqslant m\}$ is equal to $\mathbb{N}$.

7. Show that $ac+bd+ps+qr+qc+pd+qd+pc = ad+bc+pr+qs+qc+pd+qd+pc$.

8. Use the definitions of these operations in $\mathbb{Z}$ and known properties of $\mathbb{N}$.

10. Let $B = \{y \in \mathbb{Z}: 0 \leqslant y \text{ and } y+a = x, \text{ for some } x \in A\}$. Apply (P5) to show that B consists of all non-negative integers. $a \leqslant x$ implies $y + a = x$ for some y with $0 \leqslant y$. This y belongs to B, so $x \in A$.

11. Let a be a lower bound for X. Then $\{x - a: x \in X\}$ is a set of non-negative integers. Apply Theorem 1.5.

12. Apply Exercise 11 to $\{-x: x \in X\}$.

13. Apply Theorem 1.13(i) to $b - a$.

14. Use Theorem 1.13. Show that if $x \neq 0$ and $y \neq 0$, then $xy \neq 0$.

Section 1.2 (p. 26)

4. If $ab > 0$ and $a \notin \mathbb{Z}^+$, $b \notin \mathbb{Z}^+$, then $-a \in \mathbb{Z}^+$, $-b \in \mathbb{Z}^+$, $(-a)(-b) > 0$ and $a/b - (-a)/(-b)$.

5. Apply Theorem 1.16(i) to $y - x$.

6. Use Theorem 1.18, with $y = 1$. For the second part, consider first $-x$.

7. Use Theorem 1.16.

Section 1.3 (p. 42)

1. $x_n - x'_n \to 0$, $y_n - y'_n \to 0$.
 $x_n + y_n - (x'_n + y'_n) = (x_n - x'_n) + (y_n - y'_n) \to 0$.
2. $[x_n] + [y_n] = [x_n + y_n] = [y_n + x_n] = [y_n] + [x_n]$. Etc.
3. $a_n \geqslant e$ for all $n > K$. $|a_n - b_n| < \frac{1}{2}e$ for $n > L$. Hence, if $n > \max(K, L)$, $b_n > a_n - \frac{1}{2}e \geqslant e - \frac{1}{2}e = \frac{1}{2}e$.
5. Suppose that $a < b$; let $\varepsilon = b - a \in \mathbb{Q}^+$. Then $|a - b| < \varepsilon$, since $(a) \approx (b)$. Contradiction.
6. $[a] < [b]$ implies that $(b - a)$ is ultimately positive. Then there is $e \in \mathbb{Q}^+$ such that $b - a \geqslant e$. Hence, $a < b$ in $\mathbb{Q}$.
8. Let $x = [a_n]$. Then (a_n) is ultimately positive, so $a_n \geqslant e$ for some $e \in \mathbb{Q}^+$ and all sufficiently large n. Take q to be $\frac{1}{2}e$.
9. (i) By Exercise 8 there is $q \in \mathbb{Q}^+$ with $q < y - x$. Apply Theorem 1.27 to obtain $r \in \mathbb{Z}^+$ such that $x < rq$. If r is the least such, then $rq < y$ also.
 (ii) $z = x + [(y - x)/\sqrt{2}]$.
10. Apply Theorem 1.27 with $y = 1$. (See Exercise 6 on page 26.)
11. Choose $a_n \in \mathbb{Q}$ such that $|x_n - a_n| < 1/n$ for each n (by Exercise 9(i)). Then $[a_n]$ is the required limit.
12. Yes.
13. (i) The least upper bound would have to be $\sqrt{2}$, which is not rational.
 (ii) Use part (i) and Theorem 1.28.

Section 2.1 (p. 62)

2. (i) This set is finite, and listed in, say, the Oxford English Dictionary.
 (ii) One word sentences first, then two word sentences, then three word sentences, etc.
 (iii) First list all 1×1 matrices, then all 2×2 matrices, etc.
 (iv) First list all those whose entries sum to 4, then all those whose entries sum to 5, etc.
 (v) Similar to part (iv).
3. (ii) Use powers of primes.
 (iii) Use an injection $\mathbb{Z} \to \mathbb{N}$ and powers of primes.
6. $A \times A$ is countable.
7. The set of all polynomials of degree n is equinumerous with the product $\mathbb{Z} \times \cdots \times \mathbb{Z}$ ($n + 1$ factors), so is countable. Use Theorem 2.14.
9. Let $Y = A \backslash B$. Then $A \cup X = Y \cup B \cup X \sim Y \cup B = A$.
10. (i) 2^n.

(ii) If X is finite then $P(X)$ is finite. If X is countably infinite then $P(X)$ is uncountable. If X is uncountable then $P(X)$ is uncountable.

11. Let $f: A \to B$ be a bijection. Let $a \in A \backslash B$. $\{f^n(a) : n \in \mathbb{N}\}$ is an infinite countable subset of A. (It is necessary to show that its elements are all distinct.)

Section 2.2 (p. 72)

2. Let $f: B \to A$ be a bijection. Then define $g: A \cup B \to A \times \{0, 1\}$ by $g(a) = (a, 0)$ $(a \in A)$, and $g(b) = (f(b), 1)$ $(b \in B)$.

5. The result is trivial if X is empty. If X is finite and not empty, then $X \times X$ is not equinumerous with X. Suppose that X is infinite, then, and let $a, b \in X$ $(a \neq b)$, and let $f: Y \to X$ be an injection. Then define $g: X \cup Y \to X \times X$ by $g(x) = (x, a)$ $(x \in X)$, and $g(y) = (f(y), b)$ $(y \in Y \backslash X)$. g is an injection. Now use Theorem 2.18.

6. Use Theorems 2.18 and 2.19.

7. The set of all sequences of length n is just $\mathbb{R}^n$.

8. A sequence is a subset of $\mathbb{N} \times \mathbb{N}$. The set S of all infinite sequences is therefore a subset of $P(\mathbb{N} \times \mathbb{N})$. Now $\mathbb{N} \times \mathbb{N} \sim \mathbb{N}$, so $P(\mathbb{N} \times \mathbb{N}) \sim P(\mathbb{N}) \sim \mathbb{R}$. Hence $S \leqslant \mathbb{R}$. Now show that $\mathbb{R} \leqslant S$.

12. (i) $X \mapsto \mathbb{N} \backslash X$ yields a bijection onto $\{X \in P(\mathbb{N}) : X \text{ is finite}\}$.
    (ii) Similar.

Section 2.3 (p. 80)

1. If card $A \leqslant \aleph_0$ then there is an injection from A to $\mathbb{N}$, so A is countable. If A is countable and infinite then card $A = \aleph_0$.

2. Does every infinite set have a countable infinite subset?

3. Translate into terms of sets, bijections, unions and Cartesian products.

5. (i) $1 + \aleph_0 = \aleph_0 + \aleph_0$.
   (ii) $2\aleph = \aleph\aleph$.

6. Use sets and injections.

8. See Exercise 9 on page 63.

9. No.

11. See Exercise 8 on page 72.

13. (i) True. (ii) True. (iii) False. (iv) False.

Section 3.1 (p. 94)

1. (i) Yes. (ii) Yes. (iii) Yes. (iv) No. (v) No. (vi) Yes.
   (vii) No.

4. 2, 5, 16.

6. It is impossible.

9. (i) and (v), (ii) and (iv), (iii) and (vi). (Specify an open disc by a pair (x, r) in $\mathbb{R} \times \mathbb{R}^+$, where x is the x-coordinate of the centre and r is the radius.)

12. Minimal element is $\{(x, x) : x \in X\}$ (this is least).
Maximal element is $\{(x, y) : x \in X, y \in Y\}$ (this is greatest).
$\mathscr{E}$ is not necessarily totally ordered by $\subseteq$.

13. Example: $\mathbb{R}$, ordered by magnitude, and the interval $[0, 1]$, ordered by magnitude.

15. (i) and (iii).

17. $(a, b) \leqslant (x, y)$ if $a < x$ or if $a = x$ and $b \leqslant y$.

19. Let f be an order isomorphism from X to an initial segment I, and let $x \in X \backslash I$. By the first part, $xRf(x)$, but $x \in X \backslash I$ and $f(x) \in I$, so $f(x)Rx$.

Section 3.2 (p. 107)

2. $\wedge$ and $\vee$ interchange meanings. Yes.

4. The lattice of all subspaces of a given vector space (ordered by $\subseteq$). For example, the set of all straight lines through the origin and all planes through the origin, together with the origin itself and the whole of 3-space. Another example may be obtained by extending the idea of Example 3.15 by replacing the incomparable elements b, c, d by an infinite set of incomparable elements, each lying below e and above a.

7. See Exercise 2.

8. (The definition given on page 103 for maximal ideals in Boolean algebras applies to lattices in general.)
(i) Every ideal is prime.
(ii) The order here is the converse of that in Example 3.17. $\{x : 2$ does not divide $x\}$ is a prime ideal.
(v) $\{(x, y) : x \leqslant 0\}$ is a prime ideal.

9. In the lattice in Example 3.15, the set $\{a, c\}$ is an ideal which is maximal but not prime.

11. $(a \wedge b) \vee (a' \vee b') = (a \vee (a' \vee b')) \wedge (b \vee (a' \vee b'))$
$= ((a \vee a') \vee b') \wedge ((b \vee b') \vee a')$
$= (1 \vee b') \wedge (1 \vee a') = 1 \wedge 1 = 1.$
Similarly, $(a \wedge b) \wedge (a' \vee b') = 0$.

13. Let I be a prime ideal, and let J be an ideal with $I \subset J$. Suppose that $x \in J \backslash I$. Then $x \wedge x' = 0 \in I$, so $x' \in J$, since I is prime. But then $x \vee x' \in J$, i.e. $1 \in J$, so that $J = A$. Hence I is maximal. Conversely, let I be maximal and let a, b be such that $a \wedge b \in I$. Suppose that $a \notin I$ and $b \notin I$. By Remark 3.21(d), $a' \in I$ and $b' \in I$. Then $a' \vee b' \in I$, i.e. $(a \wedge b)' \in I$. But this implies that $I = A$. Hence I is prime.

Section 4.1 (p. 115)

1. (i) $\{x \in \mathbb{N}: x \text{ is even}\} \subseteq \{x \in \mathbb{N}: x \text{ is not prime}\}$.

   (ii) $\{x \in \mathbb{N}: x \text{ has a rational square root}\} = \{x \in \mathbb{N}: x \text{ is a perfect square}\}$.

   (iii) $\{x \in \mathbb{N}: x \text{ is prime}\} \subseteq \{x \in \mathbb{N}: x+2 \text{ is prime}\} \cup \{x \in \mathbb{N}: x+4 \text{ is prime}\}$.

   (iv) $\{x \in \mathbb{N}: x \text{ is odd}\} \cup \{x \in \mathbb{N}: x \text{ is even}\} = \mathbb{N}$.

   (v) $\{x \in \mathbb{N}: 3 \text{ divides } x\} \cap \{x \in \mathbb{N}: 4 \text{ divides } x\} = \{x \in \mathbb{N}: 12 \text{ divides } x\}$.

   (vi) $\{x \in \mathbb{N}: x \text{ is prime}\} \cap \{x \in \mathbb{N}: x \text{ is a perfect square}\} = \emptyset$.

2. (i) There is a y such that $y \in x$.

   (ii) There is no z such that $z \in x$ and $z \in y$.

   (iii) If there is a $u \in x$ and there is a $v \in y$ then there is a $z \in x \cup y$.

Section 4.2 (p. 128)

2. $\bigcup x = \{\{y\}, \{y, \{z\}\}\}$.

   $\bigcup\bigcup x = \{y, \{z\}\}$.

   $\bigcup\bigcup\bigcup x = y \cup \{z\}$.

4. (i) $x \cup y = \bigcup\{x, y\} = \bigcup\{y, x\}$ (by Exercise 2) $= y \cup x$.

   (ii) $x \cup (y \cup z) = \bigcup\{x, y \cup x\} = \bigcup\{x, \bigcup\{y, z\}\} = $ the set of all w such that $w \in x$ or $w \in \bigcup\{y, z\} = $ the set of all w such that $w \in x$ or $w \in y$ or $w \in z = (x \cup y) \cup z$.

5. The contrapositive is trivial. No.

7. In general $x \subseteq P(\bigcup x)$.

8. $\{x\} = \{y \in P(x): y = x\}$.

9. Let $A = \{x \in U: x \notin x\}$. A is a set, by (ZF6), so $A \in U$. If $A \in A$ then $A \notin A$, and if $A \notin A$ then $A \in A$.

11. (i) Given G, every set X with $X \sim G$ can be given a group structure isomorphic to that of G via a bijection from G. Apply Exercise 10.

    (ii) Use an argument similar to that given in Exercise 10.

12. The given set is equal to $\{y \in x: x \leqslant y\}$.

15. $\{x, y\} = \{z \in u: z = x \vee z = y\}$.

17. Let $y_n = \{x \in \mathbb{Z}: x < -n\}$, for each $n \in \mathbb{N}$.

19. A non-empty set x with the property that for each $u \in x$, $u \cup \{u\} \in x$ (as in (ZF8)) contains descending $\in$-sequences of arbitrary length.

20. Replace $\emptyset \in x$ by: $(\exists y)(y \in x \ \& \ (\forall z)(z \notin y))$. Let x be as given by (ZF8). Then by (ZF6) there is a set $\{y \in x: y \neq y\}$.

Section 4.3 (p. 144)

1. $\bigcup(x, y) = \{x, y\}$.

   $\bigcup(x \times y)$ has as elements all sets $\{u\}$ with $u \in x$ and all sets $\{u, v\}$ with $u \in x$ and $v \in y$.

4. It is $\{y \in \bigcup\bigcup R : (\exists x)((x, y) \in R)\}$. This says, in effect, that the image of a binary relation is a set, so in particular the image of a function is a set. But (ZF7) says more: it says that, for any formula determining a function, the image of a set is a set. For (ZF7) there need be no given relation. For example, the formula could be $y = \boldsymbol{P}(x)$. There is no set of all ordered pairs (x, y) with $y = \boldsymbol{P}(x)$. (If there were, its domain would be the set of all sets.)

5. Use Exercise 4.

6. f is an injection: f is a function & $(\forall x)(\forall y)(\forall z)(((x, z) \in f$ & $(y, z) \in f) \Rightarrow x = y)$.

8. R is a total order relation: $(\exists v)(R \subseteq v \times v)$ & $(\forall x)(x \in v \Rightarrow (x, x) \in R)$ & $(\forall x)(\forall y)(x \in v$ & $y \in v$ & $(x, y) \in R$ & $(y, x) \in R) \Rightarrow x = y)$ & $(\forall x)(\forall y)(\forall z)((x \in v$ & $y \in v$ & $z \in v$ & $(x, y) \in R$ & $(y, z) \in R) \Rightarrow (x, z) \in R)$ & $(\forall x)(\forall y)((x \in v$ & $y \in v) \Rightarrow (x, y) \in R \vee (y, x) \in R)$.

12. Let $A = \{n \in \omega : 0 \in n^+\}$. Clearly, $0 \in A$, since $0 \in 0^+$. Further, let $n \in A$. We must show that $0 \in (n^+)^+$. Now $(n^+)^+ = n^+ \cup \{n^+\}$ and $0 \in n^+$, so $0 \in (n^+)^+$ as required.

13. (ii) Let $A = \{m \in \omega : m + n = n + m$ for all $n \in \omega\}$.
 (iii) Let $A = \{p \in \omega : (m + n) + p = m + (n + p)$ for all $m, n \in \omega\}$.
 (iv) First show by induction that $0 \times n = 0$ for all $n \in \omega$, then by another induction that $m^+ n = mn + n$ for all $m, n \in \omega$.

14. (i) Suppose that $y \neq 0$. Then $y = z^+$ for some $z \in \omega$ (see Exercise 4 on page 18). $x + y = x + z^+ = (x + z)^+ \neq 0$ (P3*).

17. See Corollary 4.21.

Section 4.4 (p. 156)

1. (X, Y) does not exist unless both X and Y are sets, in which case it means $\{\{X\}, \{X, Y\}\}$.

2. (i) $\mathcal{A}(x)$ is $(\forall z)(z \in x \Rightarrow z \in y)$.
 (ii) Start with $\{z : (\exists x)(\exists y)(z = (x, y)$ & $\mathcal{F}(x, y))\}$.
 Now substitute another formula for $z = (x, y)$.
 (v) $x \sim y$ means there is a set f which is a bijection from x to y. See Exercise 6 on page 145. They are all proper classes except for (ii) (for certain formulas $\mathcal{F}$) and (v) (for $x = \emptyset$).

3. Let $F = \{(u, v) : u$ and v are sets & $u = v$ & $\mathcal{A}(v)\}$. F is a class, by (VNB9). Apply (VNB8).

4. Let y be a set and let X be a class. $X \cap y$ is the image of y under the function $\{(u, v) : u$ and v are sets & $u = v$ & $u \in X\}$. Yes, a class can have a proper class as a subclass.

7. It is sufficient to show that all of the axioms (including all instances of the axiom schemes) of ZF are consequences of the axioms of VNB.

Section 5.1 (p. 171)

1. The result is trivial if $x \in X$. Let Y be an infinite countable subset of X and let $Z = X \backslash Y$. $X \cup \{x\} = Z \cup Y \cup \{x\}$, a disjoint union. $Y \cup \{x\} \sim Y$, by methods of Chapter 2.

2. (i) Use Theorem 5.1.
   (ii) Use Exercise 1.

3. Use Theorem 2.18.

6. Let B be a choice set for $\{A_i : i \in I\}$. Then $B \sim I$ (using the disjointness of the sets A_i). B is a subset of A, so $I \leqslant A$.

10. For each set X in the family F, let $f(X)$ be the least element of X. f yields an element of ΠF. For $\mathbb{Z}$ use a similar (but different) criterion for choosing.

11. Let $X = \bigcap \{F(i) : i \in I\}$ and let G be the family with index set I such that $G(i) = X$ for all $i \in I$. Then $\Pi G \neq \emptyset$, and $\Pi G \leqslant \Pi F$.

12. If X were a countable infinite subset of $A \cup B$ then either $X \cap A$ or $X \cap B$ would be countable and infinite. The second part is similar.

14. If (X, R) is well-ordered then X can contain no infinite descending chain. This is immediate from the definitions. Conversely, suppose that (X, R) is totally ordered but not well-ordered. Then X contains a subset $A \neq \emptyset$ which has no least element. A must be infinite, so by Theorem 5.1 A has a countable infinite subset. This must be, or must contain, an infinite descending chain.

15. (i), (ii) See Theorem 2.31.
    (iii) It is sufficient to show that a set with cardinal number $2^{2^{\aleph_0}}$ has an infinite countable subset (without using (AC)). To do this find explicitly an infinite countable subset of $P(P(\mathbb{N}))$.

Section 5.2 (p. 183)

2. Use the standard well-ordering of $\mathbb{N}$ and an injection from the given set into $\mathbb{N}$.

3. Suppose that $A \neq X$, so that $X \backslash A \neq \emptyset$. Then $X \backslash A$ contains a least element, say x_0. All predecessors of x_0 lie in A, so by hypothesis we must have $x_0 \in A$. Contradiction.

4. The set containing the least element of each is a choice set.

5. (v) Let $\mathscr{X}$ be the set of all order relations on X such that $R \subseteq S$. By Zorn's lemma we show that $(\mathscr{X}, \subseteq)$ contains a maximal element. Then we can show that this must be a total order.

7. Apply Zorn's lemma to the set of all rationally independent subsets of $\mathbb{R}$, ordered by $\subseteq$. Then prove that a maximal element in this set spans $\mathbb{R}$, so is a Hamel basis.

9. Treat separately the possible combinations of $\kappa < \mu$, $\mu < \kappa$, $\lambda < \nu$, $\nu < \lambda$.

12. Let $\mathscr{X}$ be the set of all well-ordered subsets of X, ordered by $\subseteq$.

14. Let $P_n(X)$ denote the set of all n-element subsets of X. Then $P_n(X) \preccurlyeq X^n$. (First well-order X, then associate each n-element subset of X with an ordered n-tuple in the obvious way.) Now $X^n \sim X$, by an inductive extension of Theorem 5.16, and so card $(\bigcup\{P_n(X): n \in \mathbb{Z}^+\}) \leqslant \aleph_0 \cdot$ card X, which equals card X, by Exercise 10. Trivially card $X \leqslant$ card $(\bigcup\{P_n(X): n \in \mathbb{Z}^+\})$, so the result follows by Theorem 2.18.

15. Let B be a Hamel basis, and show that the cardinal number of the set of expressions $q_1 b_1 + q_2 b_2 + \cdots + q_n b_n$ with $b_1, \ldots, b_n \in B$ and $q_1, \ldots, q_n \in \mathbb{Q}$ ($n \in \mathbb{Z}^+$) is equal to card B.

Section 5.3 (p. 191)

1. Let $\{X_i : i \in \mathbb{N}\}$ be a set of non-empty finite sets. Let S be the set of all finite sequences (a_n) such that $a_i \in X_i$ for each i. S is certainly infinite, so by the given assumption there is an injection $g : \mathbb{N} \to S$. Now construct $h : \mathbb{N} \to \bigcup\{X_i : i \in \mathbb{N}\}$ by $h(n) =$ the nth term in the sequence $g(k_n)$, where k_n is the least number such that $g(k_n)$ is a sequence of length n or longer. Why does k_n always exist? This h is the required choice function.

2. See Theorem 5.1.

3. Let x be an accumulation point of A. For $n \in \mathbb{Z}^+$, let I_n be the interval $(x - (1/n), x + (1/n))$. The set $\{I_n \cap A : n \in \mathbb{Z}^+\}$ has a choice function f, say. Then $f(I_1 \cap A), f(I_2 \cap A), \ldots$ is a sequence of elements of A which converges to x.

4. For each $x \in A$, let $_x A$ denote the set $\{y \in A : xRy\}$. The set $\{_x A : x \in A\}$ has a choice function f, say. For each $x \in A$, denote $f(_x A)$ by $g(x)$. Then, given any $x_0 \in A$, $x_0, g(x_0), g(g(x_0)), \ldots$ is a sequence as required.

5. Let $\{X_n : n \in \mathbb{N}\}$ be a set of non-empty sets. Let S be the set of all finite sequences $(x_0, x_1, \ldots, x_n)$ such that $x_0 \in X_0, \ldots, x_n \in X_n$. Define a relation R on the set S by: sRt if and only if $s = (x_0, \ldots, x_n)$, say, and $t = (x_0, \ldots, x_n, x_{n+1})$ for some $x_{n+1} \in X_{n+1}$. Apply (DC) to S.

7. Let $\mathscr{X} = \{X_n : n \in \mathbb{N}\}$ be a countable set of non-empty subsets of $P(\mathbb{N})$. Let S consist of all infinite sequences $(n_0, n_1, \ldots)$ for which $\{n_1, n_3, \ldots\}$ (the set of numbers chosen by player 2) belongs to X_{n_0}. Player 1 cannot have a winning strategy, so player 2 does have, by (AD). We can now define a choice function f on $\mathscr{X}$ by letting $f(X_n)$ be the set $\{n_1, n_3, n_5, \ldots\}$ of numbers chosen by player 2 in response to the choices $(n, 0, 0, \ldots)$ by player 1.

Section 6.1 (p. 204)

1. ω is well-ordered by $\leqslant$. Let $n \in \omega$. Then $n = \{m \in \omega : m < n\}$ (Theorem 4.19(v)).

2. No.

3. (i) No. (ii) No. (Neither is well-ordered.)

4. Use Theorem 6.5(iv) and Theorem 2.18. The answer to both questions is no.

6. $\beta \in \bigcup \alpha^+ \Rightarrow \beta \in \gamma \in \alpha^+ \Rightarrow \beta \in \gamma \in \alpha$ or $\beta \in \gamma = \alpha \Rightarrow \beta \in \alpha$, and conversely. If α is a limit ordinal, $\beta \in \alpha \Rightarrow \beta$ is not greatest in $\alpha \Rightarrow$ there is $\gamma \in \alpha$ with $\beta \in \gamma \in \alpha \Rightarrow \beta \in \bigcup \alpha$. The converse is easy.

9. $\bigcup X$ is a limit ordinal if X has no greatest element or if the greatest element in X is a limit ordinal. Otherwise $\bigcup X$ is a successor ordinal.

10. See the proof of Theorem 6.12.

11. (i) Let $\beta = \mathrm{ord}\,(X, \subseteq)$. Then $\beta \leqslant \alpha$. If $\alpha \in \beta$, then $\alpha \subseteq \beta$ so $\alpha \leqslant \beta$, giving $\alpha = \beta$, which contradicts $\alpha \in \beta$. Hence, $\beta \leqslant \alpha$.
    (ii) Similar argument.

12. 'x is an ordinal' can be written: x is well-ordered by the relation $\subseteq$ and every element of x is equal to the set of all its predecessors. Translate this into a formula of VNB.

Section 6.2 (p. 220)

1. (i) Use the results: $\gamma < \alpha$ if and only if $\gamma \in \alpha$, and $\gamma = \alpha_\gamma$.

3. (i) Fix α. Let $A = \{\gamma \in \alpha : 0 + \gamma = \gamma\}$. Let $\beta < \alpha$ and suppose that $\beta \subseteq A$. If β is a successor ordinal, say $\beta = \delta^+$, then $\delta \in \beta$, so $\delta \in A$, i.e. $0 + \delta = \delta$. $0 + \delta^+ = (0 + \delta)^+ = \delta^+$, so $\delta^+ \in A$, i.e. $\beta \in A$. If β is a limit ordinal then $0 + \beta = \bigcup\{0 + \gamma : \gamma < \beta\} = \bigcup\{\gamma : \gamma < \beta\} = \beta$, by Corollary 6.9, so $\beta \in A$ in this case also. Hence, $A = \alpha$, by Corollary 6.16, and so $0 + \gamma = \gamma$, for all $\gamma < \alpha$. This works for any ordinal α, so the result is proved.
    (ii) and (iii) require similar arguments.

4. (iv) If X has a greatest element, say δ, then $\bigcup X = \delta$ and $\{\alpha + \beta : \beta \in X\}$ has a greatest element $\alpha + \delta$, by part (iii). Hence, $\bigcup\{\alpha + \beta : \beta \in X\} = \alpha + \delta = \alpha + \bigcup X$, using Corollary 6.9. If X has no greatest element, let $\bigcup X = \delta$, which is necessarily a limit ordinal. Then $\{\alpha + \beta : \beta \in X\}$ is cofinal in $\alpha + \delta$ (by part (iii)). By Exercise 8 on page 205, $\bigcup\{\alpha + \beta : \beta \in X\} = \bigcup(\alpha + \delta) = \alpha + \delta$ since $\alpha + \delta$ must be a limit ordinal.
    (v) Transfinite induction on γ, using part (iv).

5. As a preliminary, prove that if X is any set of ordinals then $\alpha \cdot \bigcup X = \bigcup\{\alpha\beta : \beta \in X\}$, for any ordinal α. Then use transfinite induction on γ. For a counterexample, take $\alpha = 1, \beta = 1, \gamma = \omega$.

6. First show, by induction on m, that if α is a countable ordinal then αm is a countable ordinal. Then $\omega^{n^+} = \omega^n \omega = \bigcup\{\omega^n m : m \in \mathbb{N}\}$, a union of a countable set of countable sets, under the induction hypothesis. For the second part, show first that the sum and product of two countable ordinals are countable.

Section 6.3 (p. 235)

1. This is easier than the proof of Theorem 6.31 because here we assume the existence of α_0.
2. $\bigcup\{\aleph_\gamma : \gamma < \lambda\}$ is certainly an ordinal. Denote it by α. Suppose that $\beta < \alpha$. Then $\beta \in \alpha$, so $\beta \in \aleph_\delta$ for some $\delta < \lambda$, and so $\beta \subseteq \aleph_\delta$. But $\aleph_\delta \subseteq \aleph_{\delta^+} \subseteq \alpha$, and $\aleph_{\delta^+}$ is not equinumerous with $\aleph_\delta$.
3. Derive a contradiction from the assumption that there is an aleph which is not an $\aleph_\alpha$, and the consequence that there is a least such.
6. $\aleph_\alpha \leqslant \aleph_\alpha^{\aleph_\beta} \leqslant \aleph_{\alpha+1}$. By (GCH), then, either $\aleph_\alpha^{\aleph_\beta} = \aleph_\alpha$ or $\aleph_\alpha^{\aleph_\beta} = \aleph_{\alpha+1}$. It can be shown (see the book by Drake) that the former holds if and only if $\aleph_\alpha$ contains no cofinal subset which is equinumerous with $\aleph_\beta$, and consequently for all regular cardinals $\aleph_\alpha$, if $\beta < \alpha$ then $\aleph_\alpha^{\aleph_\beta} = \aleph_\alpha$.

REFERENCES

[1] Barwise, K. J. (ed.). *Handbook of Mathematical Logic*, North-Holland, 1976.
[2] Beth, E. *The Foundations of Mathematics*, North-Holland, 1968.
[3] Birkhoff, G., & Maclane, S. *A Survey of Modern Algebra*, Macmillan, 1965.
[4] Bostock, D. *Logic and Arithmetic*: vol. 1, *Natural Numbers*, Oxford University Press, 1974.
[5] Bostock, D. *Logic and Arithmetic*: vol. 2, *Rational and Irrational Numbers*, Oxford University Press, 1979.
[6] Cohen, P. J., & Hersh, R. Non-Cantorian Set Theory, *Scientific American*, December 1967.
[7] Dedekind, R. *Essays on the Theory of Numbers*, Dover, 1963.
[8] Drake, F. R. *Set Theory*, North-Holland, 1974.
[9] Enderton, H. B. *Elements of Set Theory*, Academic Press, 1977.
[10] Fraenkel, A., Bar-Hillel, Y., & Levy, A. *Foundations of Set Theory*, North-Holland, 1973.
[11] Grattan-Guinness, I. (ed.). *From the Calculus to Set Theory 1630–1910*, Duckworth, 1980.
[12] Halmos, P. R. *Naive Set Theory*, Springer, 1974 (first published in 1960 by Van Nostrand).
[13] Hamilton, A. G. *Logic for Mathematicians*, Cambridge University Press, 1978.
[14] van Heijenoort, J. (ed.). *From Frege to Gödel: A Source Book in Mathematical Logic 1879–1931*, Harvard University Press, 1967.
[15] Hersh, R. Some Proposals for Reviving the Philosophy of Mathematics, *Advances in Mathematics*, vol. 31, no. 1, 1973.
[16] Jech, T. *The Axiom of Choice*, North-Holland, 1973.
[17] Kline, M. *Mathematical Thought from Ancient to Modern Times*, Oxford University Press, 1972.
[18] Kuratowski, K., & Mostowski, A. *Set Theory*, North-Holland, 1968.
[19] Mendelson, E. *Number Systems and the Foundations of Analysis*, Academic Press, 1973.
[20] Rubin, H., & Rubin, J. *Equivalents of the Axiom of Choice*, North-Holland, 1963.
[21] Rutherford, D. E. *Introduction to Lattice Theory*, Oliver and Boyd, 1965.
[22] Sierpinski, W. *Cardinal and Ordinal Numbers*, Polish Scientific Publishers, 1965.

[23] Stewart, I., & Tall, D. *The Foundations of Mathematics*, Oxford University Press, 1977.

[24] Stoll, R. *Introduction to Set Theory and Logic*, Freeman, 1963.

[25] Swierczkowski, S. *Sets and Numbers*, Routledge and Kegan Paul, 1972.

INDEX OF SYMBOLS

The page number given is that where the symbol is either defined or first used. Standard mathematical symbols which are used frequently and incidentally are not listed. The symbols are grouped as follows: English letters, Greek letters, Hebrew letters, mathematical symbols, logical symbols.

SUBJECT INDEX